CW00750392

MODERN CHINESE WARPLANES, Chinese Air Force – Aircraft a

Andreas Rupprecht

MODERN CHINESE WARPLANES

Chinese Air Force – Aircraft and Units

Andreas Rupprecht

HARPIA
PUBLISHING+

Consulting and inspiration by Kerstin Berger

Front cover artwork by Ugo Crisponi, AviationGraphic shows an operational J-20 from the 176th Brigade, FTTB.

Rear cover artworks by Ugo Crisponi, AviationGraphic show (from top to bottom): Y-20A '11056' from the 4th TD,
12th AR; JL-10 '78437' from the 172nd Brigade, FTTB and a Soaring Dragon II UAV from an unknown unit.

Artworks by Ugo Crisponi, AviationGraphic

Editorial by Thomas Newdick

Layout by Norbert Novak, www.media-n.at, Vienna

Maps by James Lawrence

Printed at finidr, Czech Republic

Harpia Publishing, L.L.C. is a member of

ISBN 978-09973092-6-3

Contents

Introduction

The recent reform and modernisations of the People's Liberation Army Air Force (PLAAF) are the most profound and fundamental changes, it had to face since its founding 69 years ago on 1 August 1949. Therefore, the bygone six years since the original book *Modern Chinese Warplanes* was published are a clear testimony for China's astonishing development.

Even if this 'process of great change and accelerated development' as mentioned in 2012 has widely been accepted, a lot more has happened, maybe even more than many – including myself – expected and more than one would notice at first glance. China is now deeper connected into the word-wide political and security system and most of all its military gained an increased international interest. Anyway China's affairs in regard to military aviation still remain widely unknown or generally little understood. Also a rising number aviation enthusiasts and military analysts around the world are following with increasing interest this still ongoing process that extend both to China's military ambitions and its aeronautical output. So even if during the last few years observers around the world became more and more familiar with the current operational types as well as the numerous ongoing projects, the tremendous changes, the number and diversity of modern types under development and the PLA's current ongoing and very dynamic process of restructuring nearly demanded a revised closer look onto this rapidly developing force.

For the PLAAF these changes most of all do not only mean the introduction of more modern equipment at the cost of numbers but most of all a revised structure, more modern training, strategy and doctrine. This is especially exemplified by the latest White Paper and the current issue reveals substantial content and context – often in a typical Chinese military jargon. It clearly demonstrates China's growing self-confidence and pride of an emerging global power that enjoys its growing international standing and influence; and that not only in the sense of a military power. The biggest uncertainty however – and at the same time the source of fear in the West – is a general lack of knowledge as well as understanding of this rising power or simply the 'Chinese way' to act in or with the public. As such – as M. Taylor Fravel summarises correctly – the study of a country's military doctrine and strategy is one major tool to help understanding a potential adversary's intentions as well as its potential for an eventual conflict.

This issue is even more complicated by another factor. It is not a question of whether respected analysts of the Chinese military exist and the value of their analysis. Instead, the point of contention is that quite often unreliable articles and reports of dubious content and intentions are amalgamated and presented as evidence for a certain preconceived position. This is further exacerbated not only by a lack of knowledge and certain prejudice but also due to a misunderstanding of the Chinese way of describing things. China uses a unique language, representative of its own historical genesis and political culture as well. Eventually all this is only part of yet another specialty of the 'Chinese Art of War', which is perhaps best described by an old Chinese adage inherited from Sun Tzu. Sun Tzu was a famous high-ranking military general, war strategist and tactician, but also a philosopher who is believed to have lived around the sixth century BC:

'All warfare is based on deception. Hence, when we are able to attack, we must seem unable; when using our forces, we must appear inactive; when we are near, we must make the enemy believe we are far away; when far away, we must make him believe we are near.'

'If you want to win a hundred battles, you need to understand your adversary, and you need to understand yourself.'

As such original Chinese sources are rarely clear in their content and intentions; they rarely confirm things by word *(expressis verbis)* and also official Chinese reports are often vague and leave a certain uncertainty to the reader. In consequence, all this has to be seen as part of psychological warfare in order to leave uncertainty that causes any potential opponents to under- or overestimate the PLA's present and future capability, and it also causes China's potential opponents to make funding and procurement decisions based on incomplete knowledge of China's own military developments.

For China itself, the world today is undergoing unprecedented changes, and China is at a critical stage of reform and development, not only military-wise. The 'Chinese Dream of great national rejuvenation', in which China will join with the rest of the world to maintain peace, pursue development and share prosperity and therefore China's destiny is vitally interrelated with that of the world as a whole. For China and here especially the PLAAF this means, only a strong national defence and powerful armed forces – under the strict auspice of the Chinese Communist Party (CCP) – is a guarantee for China's peaceful development. Consequently China's armed forces have to adapt themselves to changes in this new security environment, which include revised military strategic guidelines and an accelerated modernisation of the armed forces, in order to safeguard China's sovereignty, security and development interests. Also, the shift of the current economic and strategic centre of gravity ever more rapidly to the Asia-Pacific region as exemplified by the US 'rebalance' of its priorities – including military presence – means for China a new situation to its periphery in regard to security and stability.

China's answer to the current endeavours is some sort of an active defence strategy with the key element of 'anti-access/area denial' (A2/AD) capabilities; a strategy particular concerning China's currently still existing inferiority in terms of its military capabilities and weapons systems in comparison to a technological superior opponent. For China so the main goal behind this strategy is therefore not to wage a major conflict, but to present any potential opponent with the unacceptable risk of miscalculation if it interferes with what China considers a national core interest.

Concerning the content and intention of this book it is now a difficult task to manage: at first a brief historical review will be given, followed by description of the unique way to number Chinese military aircraft. Chapter three and four follow with a description of the most important operational types and weapons and chapter 5 describes the revised training syllabus and chapter 6 – in fact the centrepiece of this book – is a description of the current order of battle (ORBAT) including the latest structural changes. Finally chapter 7 concludes with a description of the airborne forces. Also since the amount of recent changes, the number of modern systems in operational use and especially the latest changes in doctrine, training and force structure are so tremendous – even more than in Naval Aviation – that they would easily surpass the available space within one volume. Consequently we decides to split the original content into three separate ones one related to the individual branches each – the Naval Aviation, the regular Air Force and the Army Aviation – and thus give each the deserved space to be described independently and comprehensively.

Again the author paid special attention to the cross-examination of the most problematic source for current developments in Chinese military matters – the Internet. Especially since President Xi gained power, much stricter and more far-ranging measures were initiated to ensure security and to restrict internet contents. Typically due to a generally vague reporting in the media and regardless the general tight secrecy and security in regards to the PRC's military, this is additionally hampered by the language barrier, that often prevents, or at least renders difficult, the process of assembling a reasonably accurate picture. Consequently the authors' standpoint holds that when it comes to the PRC and its military aviation, a lengthy learning process is required to understand the true nature of the topic in hand. In this case, quantity of information is no indicator for the reliability of a specific source: only the quality counts.

With this in mind, the author hopes that the result of this work provides an extensively illustrated compact yet comprehensive directory, with in-depth analysis of its organisation and unit structure, as well as their current equipment of modern Chinese air power. Above all, I hope that this work serves as a useful tool for many observers outside – and perhaps also inside – the PRC who are curious and eager to understand the 'rising dragon' of Chinese military aviation. While I invested considerable effort into ensuring that the sources for all illustrations presented herein are properly credited, some of these remain unknown. By the same token, the author apologise in advance for any errors or inaccuracies in this work: these are all my own.

Andreas Rupprecht, September 2018

Acknowledgements

Again, this project would not have been possible without the support of many individuals. The author would like to express his gratitude to a number of supervisors who provided invaluable assistance, support, guidance and – most of all – patience in the process of developing this project. Their deepest gratitude is offered to a number of posters on various online discussion groups especially around the internet sites China-Defense and Sino-Defence as well as their forums, without whose knowledge and assistance this work could never have been realised.

Sadly, many of them – especially those in China – prefer to remain anonymous. The author would like to thank them all for their extensive help in the provision of references, sharing of literature, translations of original publications and documentation, research into the latest reorganisations and unit insignia, as well as their unstinting moral support. One however should be named individually – Patvera – who only wanted to be credited by this nickname since he was the main help in translating original Chinese texts, explaining PLA terms and for the Chinese characters used in this book.

Last but not least, the author would like to express his gratitude to his family, for their understanding and patience throughout the duration of what was a very intensive period of work.

Abbreviations

AB	Air Brigade
AD	Air Division
ADIZ	Air Defence and Identification Zone
AESA	active electronically scanned array
AEW	airborne early warning
AEW&C	airborne early warning and control
AFAU	Air Force Aviation University
AOR	Area of Responsibility
AR	Air Regiment
ASCC	Air Standardisation Co-ordinating Committee (committee for standardisation of designations for foreign [primarily Soviet and Chinese] armament; its standardisation codenames are usually known as 'NATO designations' and have meanwhile been standardised as such)
AVIC	Aviation Industry Corporation of China
BUAA	Beijing University of Aeronautics and Astronautics (also known as Beihang University)
CAC	Chengdu Aircraft Industry Corporation (also known as CAIG Chengdu Aircraft Industry Group)
CAE	Chinese Academy of Engineering
CATIC	China National Aero-Technology Import and Export Corporation
CCP	Chinese Communist Party
CEF	Chengdu Aero-Engine Factory
CEGC	Chengdu Engine Group Company
CFTE	China Flight Test Establishment (sometimes also called Chinese Flight Test Evaluation)
CHETA	China Hai-Yang Electro-Mechanical Technology Academy
CMC	Central Military Commission
CP	command post
CUA	China United Airlines
Det	detachment
ECCM	electronic counter-countermeasures
ECM	electronic countermeasures
ECS	East China Sea
EEZ	Exclusive Economic Zone
FA	Flying Academy
FOD	forward operational deployment
FTTB	Flight Test and Training Base
FTTC	Flight Test and Training Centre
GAIC	Guizhou Aviation Industry Corporation (also known as Guizhou Aviation Aircraft Co Ltd, GAAC)
GAIG	Guizhou Aviation Industry Group
HAF	Harbin Aircraft Factory
HAIG	Hongdu Aviation Industry Group
HAMC	Harbin Aircraft Manufacturing Corporation
HOTAS	hands on throttle and stick

HQ	headquarters
i/i	in introduction
i/p	in production
KnAAPO	Komsomolsk-on-Amur Aircraft Production Association
LMC	Liyang Machinery Corporation (assigned to Liyang Aero Engine Corporation, now a subsidiary of GAIC)
MAI	Ministry of Aircraft Industry
MID	Mechanical Industry Department
MR	Military Region
MRAF	Military Region Air Force
MRTB	Military Region Training Base
n/a	no information available
NA	Naval Aviation (People's Liberation Army Naval Air Force, PLANAF, is an unofficial term, most often used in the US)
NAD	Naval Aviation Division
NAMC	Nanchang Aircraft Manufacturing Company
NAU	Naval Aviation University
n/k	not known
NPU	Northwestern Polytechnical University
NRIET	Nanjing Research Institute of Defence Technology
NUDT	National University of Defence Technology
ORBAT	order of battle
PESA	passive electronically scanned array
PLA	People's Liberation Army
PLAAA	People's Liberation Army Army Aviation
PLAAF	People's Liberation Army Air Force
PLAN	People's Liberation Army Navy
PLANAF	(see NA, Naval Aviation)
PRC	People's Republic of China
ROC	Republic of China (Taiwan)
RoCAF	Republic of China Air Force
SAC	Shenyang Aircraft Industry Corporation
SADO	Shenyang Aero-engine Design Office
SAEF	Shenyang Aero Engine Factory (now Shenyang Liming Aero Engine Company – Liming Engine Manufacturing Corporation)
SAIC	Shaanxi Aircraft Industry Corporation (also known as Shaanxi Aircraft Corporation)
SCS	South China Sea
SEF	Shenyang Aero-Engine Factory
TC	Theater Command
TCAF	Theater Command Air Force
TR	Training Regiment
TTC	Test and Training Centre
XAC	Xi'an Aircraft Industrial Corporation (also known as Xi'an Aircraft Company)
XAE	Xi'an Aero-Engine Corporation

HISTORY AND FUTURE OF THE CHINESE AIR FORCE

CHINESE AIR FORCE

To understand the latest developments and the current issues China and its armed forces are facing, one has to look back to the roots and the history of this fascinating air force. By doing so, and by using Chinese sources – and the most authoritative Western publications – the People's Liberation Army Air Force's (PLAAF) history can be traced through five periods, or six if current developments are included.

Prequel: 1924–49

In terms of accurate assessments, almost all attempts to describe the development of military aviation in China over the first half of the 20th century have proved to be incomplete. The utility and importance of aviation was quick to be recognised in China, and the development of military aviation had begun as early 1910. However, until a semblance of unity was brought about during the Sino-Japanese War, China existed in what was practically a state of chaos, characterised by incessant quarrelling between petty factions and warlords, many of whom established and maintained their own air services. Even after the establishment of the central government in 1924, the progress made towards a more stable regime remained limited. Significantly, the government did not control China in its entirety, and much of the modern weaponry acquired to counter the growing threat from Japan was lost in the course of the prolonged campaigns waged against the Chinese Communist Party (CCP) and various warlords. The situation only worsened following the Japanese invasion of Manchuria in 1931-32, when the US, Italy, the USSR and other parties became involved on the side of major Nationalist and Communist factions. To make the situation even more complex, in the following years the Japanese authorities created several states within China, and some of these began establishing their own air arms. At the same time and following a split between the Chinese Nationalists and the CCP, the latter began establishing its own military: this also included efforts to create what became a rather short-lived air force.

Over the next 15 years, US efforts in particular, resulted in the unification of some of the existing military air services into the Chinese Air Force (CAF), run by the Nationalists. The CAF underwent wholesale reorganisation and re-equipment, predominantly receiving US-made combat aircraft and transports. On the cessation of hostilities with Japan in September 1945, and in addition to an existing US-supported air arm

The majority of aircraft initially operated by the PLAAF were of US origin, as evidenced by this photograph from 1949, showing several North American P-51 Mustang fighters. (CDF)

of around 300 aircraft, the Chinese government purchased the entire stock of surplus US aircraft in the Chinese Theater. This amounted to approximately 1,000 aircraft and these were further increased by more than 200 others from Canada and the UK. Due to a shortage of trained personnel, however, and the continued defection of air force crews to the advancing Communist forces, the CAF was incapable of operating more than 200 of these aircraft at any one time.

The years 1946-47 saw a rising tide of Communist successes, and in January 1949 the Nationalist government began to decline rapidly as the People's Liberation Army of the CCP occupied province after province. By January 1950 the Nationalists only retained control of several islands along the Chinese coast, the largest of which were Formosa (Taiwan) and Hainan. Over the following eight years, the Nationalists were to lose most of these in a series of military campaigns launched by the Communists, and they would eventually entrench themselves on Taiwan. Here, the remnants of the CAF were reorganised as the Republic of China Air Force. Remaining as an entirely separate service to this day, the history of the RoCAF is beyond the scope of this book.

Meanwhile, on the Communist side, while the few aircraft captured from the Nationalists in the 1930s were not sufficient to establish an air force, their acquisition is now often regarded as marking 'the first beginnings' of an air arm of the People's Republic of China. Instead, it was only when the CCP's Central Military Commission (CMC) had established the Air Force Engineering School in January 1941 that serious efforts in this direction were launched. The Air Force Engineering School lacked aircraft and airfields, but was run by two CCP cadets, Wang Bi and Chang Qiankun, who had previously served in the Soviet Air Force. After slow beginnings, Communist efforts to establish their own air force were continued in May 1944, when the CMC set up a subordinate Aviation Section with Wang and Chang as the director and deputy director respectively. Almost exactly two years later, the Northeast Old Aviation School was established in Mudanjiang, Jilin Province. This unit was equipped with four basic trainers and a few advanced trainers and also included a large number of former Imperial Japanese Air Force personnel who remained in China following the end of World War II.

During 1949 and 1950 the Communists captured hundreds of CAF aircraft, 115 of which were in operational condition. In March 1949, the CMC therefore upgraded its Aviation Section to an Aviation Bureau, with Chang as the director and Wang as the political commissar. Finally, on 11 November 1949, the CMC disestablished the Aviation Bureau and officially established the People's Liberation Army Air Force.

Foundations: January 1949 to December 1953

China and the Chinese are rooted in thousands of years of rich culture and history. It was therefore hardly surprising that during the first few years of the CCP's rule, the new government reorganised almost all aspects of Chinese life. Measures were soon adopted to curb inflation, restore communications and re-establish the order necessary for economic development. At the same time, campaigns were orchestrated to mobilise mass revolutionary fervour, and to remove from power those likely to obstruct the new government. The results of such programmes included fundamental changes within society and the military. The history of the PLAAF during its founding period was no exception to these trends.

In the mid-1950s, the majority of fighter divisions were equipped with MiG-17 and licence-manufactured J-5 fighters. Both played an important role during operations against islands off the coast held by the Chinese Nationalists. (CDF)

Originally established from elements of the People's Liberation Army (PLA), the fledgling air arm was to maintain close connections to the new political system via the CMC and it would operate under a very similar organisational structure to that of the PLA. At its head was the highly valued 'dual command system', consisting of a commander of proven experience and reliability – Liu Yalou – and a similarly seasoned political officer, Xiao Hua. Similarly, each of the PLA's 13 original Military Regions (MRs) established its own Air Division with subordinated Aviation Offices, each of which was under the command of its own commander and a political officer. Over time, the number of MRs was reduced, so that by 1970 there were 11 and in 1985 seven were reorganised into Military Region Air Forces (MRAFs).

The strong influence of the PLA's experiences could be found in almost every other aspect related to the PLAAF, including its organisation. So strong was this tendency that the leadership of the CCP and the CMC expected the experienced PLA fighters to become accomplished airmen almost as fast as they learned to use rifles, artillery and armour. Under such circumstances, one of the early objectives of the PLAAF and its MRs was to take over as much as possible former CAF airfields and aircraft, maintenance and manufacturing facilities, and also trained personnel. Another parallel to the history of the PLA can be found in the fact that when the PLAAF established its first provisional unit, in July 1949, this was designated as the 4th Mixed Air Brigade, which was later expanded into a full division, to commemorate the PLA's 4th Army that had fought in the 1930s.

For many reasons, the subsequent expansion of the PLAAF proceeded very quickly. The CCP was not only interested in consolidating itself in power and recovering the economy as soon as possible – in part through the development of a technologically sophisticated air force – but also in planning the liberation of all the islands still held by the Nationalists, particularly Taiwan. In order to further accelerate the formation of their air force and to gain additional expertise, the Chinese Communists established close ties with the relevant authorities in the USSR and placed sizeable orders for aircraft and equipment as early as mid-1949. Furthermore, in response to Nationalist air raids in the Shanghai area, several units of the Soviet Air Force (VVS) were deployed to China during the spring of 1950.

In June 1950, North Korean forces invaded South Korea, sparking a major war against UN forces and which was to continue for the next three years. The outbreak of the war took the command of the PLAAF by surprise, and with the air arm unprepared for a large-scale conflict. The Air Force had been established barely eight months earlier and was still in the process of organising its first two fighter divisions, one bomber regiment, and one attack regiment, which operated a total of around 200 combat aircraft. Most PLAAF pilots had accumulated fewer than 100 flying hours and most MiG pilots still hadn't flown their first solo sorties. Ultimately, the original primary mission of the PLAAF was to provide direct support to the PLA in the re-conquest of islands still held by the Nationalists. The unexpected outbreak of the war in Korea distracted the leadership of the CCP and the PLAAF from preparations for such operations. Instead, the PLA was ordered to enter the 'War to resist Americans and aid Korea' (as the Korean War is officially known in China). Meanwhile, the PLAAF was forced to gear-up to counter the highly experienced UN and US air arms and provide direct support to PLA volunteers. For these purposes, the PLAAF contigent sent to Korea was designated as the Chinese People's Volunteer Air Force.

Due to a number of understandable difficulties and following extensive preparations in the form of training the necessary personnel, and establishing the required logistics and aircraft maintenance support, a small group of PLAAF MiG-15 pilots began flying combat missions over Korea under Soviet supervision in December 1950. Unsurprisingly, a lack of experience and available assets prevented large-scale operations before autumn the following year. Additionally, the situation on the ground in Korea had changed significanty, resulting in the realisation by the leadership in Beijing that a protracted war resulting in stalemate was now more likely than the swift Communist victory originally expected. Therefore, new strategies were adopted among which was a plan to have the PLAAF gradually rotate its units in and out of the theatre of operations, so as to offer its novice airmen the opportunity to gain combat experience and improve the methods of command. Therefore, although no fewer than 12 PLAAF air divisions equipped with MiG-15s and bombers were soon available for operations over Korea, it was only during early 1952 that the Air Force became involved in large-scale operations. From that time onwards, pressurised by the leadership in Beijing to rotate as many units in and out of the conflict zone and thus obtain as much experience as possible, the PLAAF became ever deeper involved in the Korean War. Coupled with this was Beijing's obsession with the idea of striking back at the enemy using its own air power, instead of being on the receiving end of unchallenged air attacks. In the process, the PLAAF learned many valuable lessons which helped its leadership continue working towards its expansion and improvement.

Although many of the lessons the PLA and the PLAAF learned during the Korean War were subsequently ignored or lost due to a series of domestic upheavals during the 1960s, the Chinese nonetheless drew many useful conclusions. Before Korea, most PLAAF pilots and officers were combat-tested infantry soldiers, without any kind of knowledge of aerial warfare. By 1953 the PLAAF had been expanded through the establishment of 13 aviation schools which had trained 5,645 flight crew and more than 24,000 maintenance personnel by 1953. In addition, it had formed 27 or 28 air divisions (including between 56 and 70 regiments), with around 3,000 aircraft. Roughly a third of these assets were deployed in air operations over Korea, yielding mature combat pilots and preparing this force for future conflicts. The Chinese learned not to underestimate their opponents, and in particular to respect the flexibility, firepower and the will to learn from the US military. At an operational level, the PLAAF understood that the PLA's traditional operations and tactics, based as they were on the annihilation of complete enemy units, did not work. The Air Force also realised the importance of professionalism, the role of firepower and logistical preparations, and adopted more cautious and realistic strategies. As well as increasing the size of its own forces, it would focus on obtaining operational experience. At the tactical level, the Chinese realised that the use of World War II-era Soviet tactics and flying in massed formations were too rigid and inflexible. The Chinese evaluated US air tactics and concluded that the Americans were more effective since, for example, they flew in pairs, or in four- to eight-ship formations, with fighters distributed at different altitudes for supportive interaction. Correspondingly, PLAAF fighter pilots soon began developing their own tactics in which they operated in 'attack and cover' pairs in order to improve their flexibility and safety. It was in this fashion that China came to possess the third largest air force in the world – and one of most combat-experienced – rivalled in terms of size only by those of the US and the USSR.

Expansion: January 1954 to April 1966

After the end of the Korean War, the CCP again turned its attention towards tackling the remaining Nationalist strongholds off the coast of mainland China. Attempting to avoid any kind of large-scale campaign or provoke the US so soon after the end of the Korean War, the leadership in Beijing decided to first attack the Dachen and Nanchi/Yijiangshan islands, off the coast of Fujian Province, some 200km (124 miles) north of Taiwan. This time the PLAAF was assigned the task of performing reconnaissance, attack, fighter escort and air defence operations, and was involved right from the start. After sufficient forces – a total of five PLAAF and three Naval Aviation divisions with around 200 combat aircraft – had been concentrated in Fujian Province, where the Nationalist air arm was still in control of the airspace, operations in the area began in March 1954. Initially, the PLAAF and the Naval Aviation fought a number of air combats in order to secure air superiority over Fujian Province. During the second phase, beginning in November 1954, Communist Chinese air power took part in a carefully conducted campaign aimed at destroying the defences of Yijiangshan, but without provoking the units of the US Navy deployed nearby. By January 1955 the PLAAF and Naval Aviation were in complete control of the airspace around Yijiangshan and they began a campaign of pounding local Nationalist positions. The garrison was eventually overrun following an amphibious landing on 25 January 1955.

Following the initial operational use of Il-28s in the First Taiwan Crisis 1955, the PLAAF introduced several more licence-manufactured H-5 bombers from 1967 on.
(Top.81 Forum)

The relatively brief but highly successful Yijiangshan campaign became a formative experience in the history of the PLAAF and the Naval Aviation. Not only was it the first operation of its kind in which the Air Force had played such an important role, but it also provided Chinese airmen with valuable experience concerning coordinating aerial and ground operations, achieving and maintaining air superiority and organising and planning air attacks, as well as the necessity of allowing the pilots involved a degree of flexibility in target selection. Indeed, this campaign proved so successful that the Nationalists were subsequently forced to accept a US proposal and evacuate Dachen Island, too.

Following this experience, the CCP ordered the PLA to begin planning for a similar operation against Quemoy and Matsu Islands. In order to be able to participate, the PLAAF first had to establish a number of new airfields in the area adjacent to the Formosa Straits and then re-deploy its units there. Once again, the PLAAF would have to establish air superiority over the Nationalist air arm in order to enable freedom of operations for its own reconnaissance aircraft and bombers.

The preparations for this new offensive were completed during the summer of 1958 and were quickly realised. Within only three weeks the PLAAF had successfully established a network of airfields and an entire air defence system capable of supporting large-scale operations adjacent to Taiwan. In the course of this phase, the PLAAF combined the units of the newly established 1st Air Corps from Fuzhou with those of the 5th Air Corps from Hanghzou to establish the Fuzhou Military Region Air Force. The result consisted of five complete and four incomplete fighter divisions (all equipped with MiG-15s, but some also flying newly delivered MiG-17s), two bomber regiments and a part of the Naval Aviation's 4th Naval Air Division, totalling 620 combat aircraft. All the aircraft, tens of thousands of tons of the necessary support equipment and fuel, thousands of personnel as well as the complete command structure were also moved to new bases.

The planning for the coming operation initially concentrated on establishing air superiority and suppressing the Nationalist forces on Jinmen and Matsu Islands, with the Chinese paying special attention to avoiding an escalation that could lead to a war with the US – which had meanwhile signed a defence pact with Taiwan. Correspondingly, PLAAF and Naval Aviation pilots received very strict rules of engagement (RoEs). These dictated that there were to be no operations over the high seas, no bombing of Matsu and Quemoy unless the Nationalists bombed the mainland, and no engagements with US aircraft unless they violated Chinese airspace. Phase one of the attack on Jinmen began on 27 July 1958, with the PLAAF dispatching large formations of MiG-15s and MiG-17s over the island, provoking a number of air combats with the Nationalist air arm. Additional formations attempted to intercept Nationalist reconnaissance fighters, as well as fighter-bombers which attempted to support the Quemoy garrison. The appearance of the new and considerably more powerful MiG-17 caused some surprise on the Nationalist side, and initially the air superiority began to tilt towards the Communist side. This enabled the PLA ground forces to deploy in positions from which they could put Jinmen under artillery fire. With this achieved, the second phase of the operation opened on 23 August.

However, by this time the US Marine Corps had rushed a batch of newly developed AIM-9B Sidewinder air-to-air missiles to Taiwan (only months after this weapon had been introduced into service in the US) and a team of specialists helped install these on the North American F-86F Sabre fighters of the Nationalist air arm. Furthermore, the US Air Force deployed 144 North American F-100 Super Sabres and Lockheed F-104 Starfighters to Taiwan and Okinawa, while the US Navy deployed no fewer than six aircraft carriers and an armada of more than 150 escort and support ships and 500 combat aircraft in the Formosa Straits. With US fighters flying defensive combat air patrols over Taiwan, the Nationalists were free to operate in force over the mainland, and thus they were not only able to prevent the PLAAF from establishing air superiority over Jinmen and Matsu, but at the same time remained capable of re-supplying besieged garrisons. Following a series of bitter air battles during September and October 1958, in which both sides suffered a number of losses, and after experiencing several command and operational problems and deficiencies in radar coverage, the PLAAF found itself in a defensive posture. As such, it had to depend upon ground control to initiate timely operations by its interceptors. Since the available radar network and the coordination between pilots and ground forces proved unsatisfactory – and facing the massive presence of technically superior Nationalist fighters in the air – the Communists failed to establish air superiority. Additionally, the Communist leaders were forced to realise that the US assistance provided to the Nationalists made an invasion of either Jinmen or Matsu, not to mention Taiwan, extremely risky in regard to a possible escalation into an open war with the Americans. Therefore, Beijing eventually decided to de-escalate the hostilities, even though Jinmen was subject to constant artillery barrages, many of which were designed to serve as a 'means of communication with the enemy', for years to come.

In this period, PLAAF expansion was dictated mainly by the lessons of the US air attacks in Korea and the threat posed by American bombers armed with nuclear weapons. Correspondingly, the Air Force emphasised the development of air defence capabilities, with fighter-interceptors constituting its largest and most important element. The expansion of the bomber fleet remained limited to achieving the capability

to support ground and naval forces. The USSR continued to strongly influence PLAAF expansion, providing it with aircraft and helping structure its organisation. While the number of air divisions remained stable at 28, with Soviet help the number of PLAAF training establishments increased to 29. More importantly, and in addition to developing their own combat tactics, over time the Chinese began drafting their own regulations and teaching materials, based on their own experiences. Previously, they had relied on translated Soviet documentation and instructional materials. Furthermore, in 1957 the PLA decided to re-integrate its Air Defence Force – originally created in 1948-49 – within the PLAAF, adopting a system in which air operations were combined with land-based air defence. During the following year, the PLAAF's ground-based air defence units were equipped with their first SA-2 SAM systems acquired from the USSR. However, the end of this period of development was characterised by the break in relations between the PRC and the USSR, between 1959 and 1961. This resulted in the near termination of work on a number of projects related to the PLAAF, foremost among them the efforts to establish the manufacturing industry for various advanced Soviet-designed combat aircraft types. Therefore, as of 1958–59, the PLAAF was one of the most modern air arms anywhere in the world but it was to enter its next phase in an incomplete condition.

Cultural Revolution: 1966 to 1976

By April 1964 the Chinese Communist leadership had become increasingly concerned about a possible all-out war with the USSR on one side, and a possible war with the US over Vietnam on the other. Both the PLAAF and Naval Aviation had become involved in the Vietnam War. They deployed their air defence assets in the north of that country and also intercepted US combat and reconnaissance aircraft which penetrated the airspace over southeast China. Concerned about a possible escalation, Chinese pilots were initially ordered only to monitor US aircraft penetrating their air space. However, when, by 1965, the number of such violations had increased, Beijing felt forced to make public an air defence demarcation line at the border. By doing so it was hoped to deprive the Americans of any excuse for what were seen as provocative over-flights. Chinese fighter pilots deployed at air bases in Guangxi, Hainan Island, the Leizhou Peninsula and Yunnan received permission to open fire. Over the following year, they claimed to have shot down 12 and damaged four US aircraft. Additionally, the PLAAF and Naval Aviation established a number of special units organised and equipped to intercept US unmanned aerial vehicles (UAVs) that were deployed against China. By the end of 1965, no fewer than 20 such drones were reported as shot down.

Throughout 1966, the US administration expanded the list of targets within North Vietnam which could be attacked by its air power with the result that Beijing felt forced to bolster its southern and southeastern frontiers. Nevertheless, the PLAAF never openly intervened on the Vietnamese side during the war: indeed, PLAAF and Naval Aviation pilots received very strict RoEs which prohibited any kind of operations over Vietnam. Therefore, related Chinese air operations remained strictly limited to the protection of the PRC's airspace during 1967 and beyond. Subsequently, the CCP leadership introduced measures intended to permit an approach to Washington in order to win the US as an ally in its own struggle against the USSR. Throughout this

The political decision to change the manufacturing site from Harbin to Xi'an delayed introduction of the H-6 for many years. Serial H-6A bombers were only delivered from 1969 onwards. (FYJS Forum)

period, small-scale aerial confrontations with the Nationalist air arm occurred time and again, and several Nationalist reconnaissance aircraft were shot down, some of them while flying deep over mainland China.

The PLAAF continued to acquire new and advanced aircraft and weaponry during the early 1960s, significantly the MiG-19S, and its locally manufactured J-6 variant, followed by the MiG-21F-13 together with R-3S air-to-air missiles. New equipment enabled the formation of no fewer than 22 additional air divisions throughout China, making a total of 50 by 1970 – the greatest number in its history. However, in the meantime China had entered the period of the Great Proletarian Cultural Revolution, commonly known as the Cultural Revolution. This was a socio-political movement set in motion by Mao Zedong. Chairman of the CCP since 1934, Mao was the architect and founding father of the PRC although he had lost much of his political influence over time. Instigating a process of 'controlled anarchy' to remove his inner-party opponents, Mao's new movement led to widespread chaos and the disintegration of all levels of government, together with the evaporation of social norms. The Cultural Revolution badly damaged China in economic and social terms, and also involved the PLAAF. The Air Force C-in-C at the time, Wu Faxian, became embroiled with Defence Minister Lin Biao, who was allegedly involved in a coup attempt against Mao in September 1971. Lin apparently fled in a PLAAF aircraft which then crashed in Mongolia. Subsequently, the CCP accused Wu Faxian of complicity, arrested him, tried him 10 years later, and sentenced him to 17 years in prison. Under the chaotic conditions prevailing in China in those days, the PLAAF did not receive a new C-in-C until May 1973, when the CMC appointed Ma Ning, former Deputy Commander of Lanzhou MRAF, to the position. Although the PLAAF reached its historical peak in terms of personnel strength – estimated at 760,000 by 1972 – the lack of guidance from Beijing, diminishing political trust, shortages of fuel and spare parts, and maintenance problems related to the widespread chaos, combined to result in serious setbacks. These manifested themselves particularly within training and educational institutions, while pilots and ground personnel had their flight training and exercises significantly reduced. Subsequently, the PLAAF was gradually reduced in size.

Modernisation: 1976 to the present

Throughout the late 1970s and early 1980s, the PLAAF slowly began to implement readjustments with regard to leadership, training, combat readiness and other aspects of peacetime operations. However, no significant improvement had occurred before 1985, when Wang Hai was appointed C-in-C PLAAF. Hai initially concentrated on improving the training and education of his personnel, and then on acquiring new aircraft and equipment based on advanced Western technologies. Meanwhile, operational-level doctrine began shifting primarily from being able to provide an adequate air defence capability for major cities and industrial areas, towards the goal of being prepared for simultaneous offensive and defensive operations.

During the late 1980s, China established various levels of cooperation with the US and several countries in Western Europe, and the PLAAF could look forward to boosting its capabilities through the introduction to service of correspondingly modified, locally manufactured aircraft types, or even imported aircraft and technology. How-

Developed during the late 1980s, the Q-5D was for many years the most capable variant of this indigenously developed type. Uniquely, it always wore a green camouflage scheme. (Top.81 Forum)

ever, most of the related projects came to a sudden end in 1989, when such ties were cancelled due to the unrest in Beijing and the massacre of demonstrators at Tiananmen Square.

Around the same time the Chinese began also to study the changing nature of modern air warfare in small wars. Ultimately, the US-led operations during the war with Iraq in 1991 shocked the PLA into the realisation that it had to become capable of engaging in high-tech warfare or otherwise face the certainty of falling ever further behind other modern militaries. The results of this shock had a galvanising effect on Chinese military leaders. As early as 1993 the leadership of the CCP and the PLA issued 'The Military Strategic Guidelines for the New Period' – equivalent to a new national military strategy. The objective of this was rapid modernisation in order to enable the PLA to fight and win wars based on high-tech weapons, joint operational concepts and high-tempo operations. Ever since, the PLA has been engaged in a long-term effort to transform itself into a force that can win modern wars but also engage in non-traditional security operations. The PLA remains in a process of sustained reform and modernisation, such developments often taking place simultaneously across a wide front and in myriad endeavours. At the same time, the PLA faces many challenges and has to cope with a wide range of inevitable dislocations and uncertainties. Nevertheless, the fact that this process continues to this day is evidenced by the introduction of additional measures over the years. Most recently, such efforts have resulted in the appearance and introduction into service of some of the most advanced fighter aircraft anywhere in the world, including the indigenous Sukhoi Su-27 and Su-30 developments such as the J-11 and J-16, the J-10 and most significantly the J-20.

Latest developments and future prospects

The periods describing the PLAAF's history often end with the epoch 'the future'. However, given the latest revolutions – which are no longer merely evolutional steps – it might be worth adding a sixth period. This began with the reforms initiated in 2012, and entered a second level in 2015 with the official unveiling of China's first white paper on military strategy on 26 May 2015 in Beijing. For the PLAAF, the current reforms mean that for the first time in its history it will face the necessity of shifting its focus from primarily territorial air defence to the ability to conduct offensive and defensive operations as well, and which can meet the requirements of informationised operations. This inevitably includes – and in public this is naturally what receives the most attention – the introduction of new hardware such as modern multi-role-capable combat aircraft. But probably the most important reforms are largely unnoticed and these concern the PLAAF's tactics and training and its command structure. Here the main focus lies currently in the proper use of these latest assets in order to better utilise their full potential and conduct different aerial combat missions which are no longer in a pre-exercised or directed manner through exercises with an increasing complexity of simulated war scenarios. Also, the complexity of these latest exercises has been expanded in order to improve coordination between different units of the PLAAF but also in joint PLAAF/Naval Aviation exercises. One major component of this reform was therefore the transformation of the former seven Military Regions and their commands into the now established five Theater Commands in parallel with the

Since September 2017 and January 2018 two J-20A prototypes, 2021 and 2022, have been flying with indigenous turbofan engines allegedly called WS-10C and featuring a LOAN-type nozzle with sawtooth edges. This suggests that the PLAAF intends to use this advanced engine instead of its Russian counterpart. (SDF)

introduction of the base/brigade concept which replaces the former division/regiment structure.

Consequently, the future of the PLAAF now appears bright and it can be summarised that within the past decade, China's military – both in strategic considerations and the modernisation of procedures and material – has undergone some of the most profound reforms since its establishment, even though some aspects might at the present time appear unclear. Other important shifts are the focus on improved training, education and tactics and close cooperation between the two air arms, whether on the tactical level or in terms of procurement, training and so on. Such developments would have been almost unheard of in the past and it remains to be seen to what extent the PLAAF will be affected by this overall military reform and how their units will be integrated into this new security system.

PLAAF AIRCRAFT MARKINGS – SERIAL NUMBER SYSTEM

NATIONAL MARKINGS

Through their long historical genesis, Chinese military aircraft wore a plethora of different national markings, especially in the first half of the 20th century, when no fewer than 27 different 'air forces' existed in China, and each of them employed slightly different sets of markings. Since 1949, following the founding of the PLAAF, all PLA military aircraft have been marked with the traditional 'star and bar' marking, applied in bright red and outlined in yellow. This device is usually located prominently on each side of the rear fuselage or on either side of the tail fin and on the upper and lower surfaces of both wings. The central star motif contains the Chinese digits '8' over '1', applied in yellow, indicating 1 August (1927), the date of the formation of the PLA. Despite the apparent standardisation of the star and bar marking, at least three different basic versions have been observed to date, as well as four different methods of the application of the markings. This changed slightly in mid-2016, when the PLAAF introduced, for the first time, toned-down markings on its first J-20A LRIP aircraft and since early 2018 on the J-16s. Consequently, a wider use could be expected in the near future.

PLAAF flag

PLA roundel

PLA low-visibility roundel

PLAAF serial number system

One of the most striking external characteristics of all Chinese aircraft are their large serial numbers, usually consisting of five digits. In the past these were often applied in a stencilled style, although more recently four 'solid' variants have become increasingly common. During the turbulent 1950s and 1960s, aircraft of the Chinese Air Force and Naval Aviation were marked only with very simple, two-digit serial numbers, usually applied in dark red, but also in black, blue or yellow. Many of these had been applied in Soviet factories, and some while the aircraft served with Soviet units deployed in China. At least as often, and particularly during the 1960s, PLAAF and Naval Aviation aircraft wore various political slogans on the forward or centre fuselage, or on other parts of the airframe. A first indigenous serial number system was not introduced until the late 1950s, and consisted of four digits, based on a coded unit designation and the aircraft's assignment to a specific regiment within the division. The current situation has since become far less easy to understand – and in contrast to 2012 – there are currently five different serial systems in use. These are as follows:

Pilot shoulder patch with PLAAF emblem

The huge yellow serial numbers on this Y-20A assigned to the 4th Transport Division are a prime example of the old serial number system that is still in use with the bomber, special mission and transport divisions. (acer31 via SDF)

The first one has its legacy in the 1970s, when at some point the PLAAF introduced a system based on five digits. This system is based on a coded divisional designation combined with the aircraft's individual number, which in turn depends on the aircraft's assignment to a specific regiment within the division. This system received a first modification – in fact it was streamlined to a common system – in mid-2005, when the PLAAF introduced a slightly simplified variant, which is best described as follows: the first and fourth digits are the relevant ones to identify the division and since 2005 it has always been aa minus 11 equals division. The second, third and fifth digits denote the allocation of the aircraft to a specific regiment and squadron since each of the regiments within the same division is allocated individual aircraft numbers. Here the numbers range from 001 to 049 for the first regiment, from 051 to 099 for the second and from 101 to 150 for the third regiment. Usually for day-to-day service the third and fifth digits represent the individual aircraft's number within a regiment. This system limited the number of possible serial numbers per regiment or brigade to 50, indicating that the PLAAF does not expect any of its major units to operate more aircraft than this. Alternatively, this system can be described as follows:

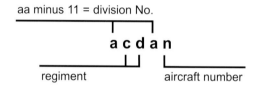

This system is currently in service within the 'large aircraft' units like the bomber, special mission and transport divisions. In parallel to this system several aircraft have long been flown by independent regiments and training units using four-digit numbers but since most of these aircraft have been renumbered it is not described in detail.

The second system was introduced in line with the process of reorganisation and restructuring which began in the second quarter of 2012. This system remains in use and in it the third and the fifth digits signify the individual number of an aircraft within the brigade, while the first, second and fourth digit minus 611 produces the number of the brigade. Confusingly, this system results in numbers within the 6xxxx and 7xxxx series – numbers which were previously allocated to former MR training bases (which were in due course disbanded) and flying schools, which are also undergoing restructuring. Similar to the original system after 2005, the third and fifth digits represent the

This image demonstrates the new serial system introduced in 2012 for air brigades – seen here is a J-10B (foreground) and Su-30MKK (background). Consequently, most combat brigades start with the numbers 6 or 7.
(Longshi via PDF)

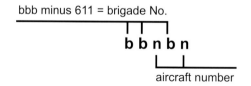

individual aircraft's number within a brigade and this system is now (2018) valid for most tactical assets including fighter, fighter-bomber and ground-attack brigades.

The third system in use, which is still not fully understood, is reserved for 'Theater-subordinated' units including reconnaissance, SAR, survey and transport brigades. As such, they function in a similar role to the former Military Region liaison units now known as Theater Command brigades. In this system the second digit runs from 1 to 5 in protocol order starting with the Eastern TC (1), Southern TC (2), Western TC (3), Northern TC (4) and the Central TC (5). It is probable that the third digit represents the type of helicopter, namely either the Mi-171 and Z-9 (6) or the Z-8K (7). If so, the final two digits would then represent the aircraft's individual number.

The fourth system is a return to a system similar to the original four-digit serial numbers previously used by liaison, transport and training units and has now been commonly reintroduced within all Flight Colleges. Known as AxBx numbers where the first digit A denotes the Flight College: Harbin (1), Shijiazhuang (2) and Xi'an (3). The third digit B denotes the individual brigade within that Flight College and therefore ranges from 1 to 5. Finally, the numbers xx are reserved to identify the individual aircraft within a certain brigade of a given Flight College.

The fifth and final system is reserved for the transports and helicopters operated by the Airborne Forces and is in fact a relic of the original four-digit serial numbers previously used by liaison, transport and training units. Until early in 2017 all helicopters used 6x6x numbers – standing for the 6th Transport Aviation Regiment, 15th Airborne Corps – but from 28 April 2017 the Airborne Forces reorganised these numbers, changing them to 6xAx. The exact meaning of the third digit A is unclear but it might differ between aircraft (6x1x) and helicopters (6x2x). The second digit denotes the type operated and the following numbers for the following types have so far been identified: Y-5C, Y-12D and Y-8 (1) and Z-8K (0), Z-9 (1) and Z-10K (2 and 3). Given the fact that nearly all services replace the former regimental numbers of a four-digit system with the now standard brigade system, it is most likely that the Airborne Forces will also follow suit and introduce the five-digit system, perhaps similar to the one recently introduced by the 'Theater-subordinated' units.

In summary the current five-digit serial numbering system can be described as follows:

- All serial numbers beginning with 1, 2, 3, and 4 – apart from those for regiments which are not converted to brigades – are for 'large' aircraft and strategic assets such as bomber (8th, 10th and 36th Divisions), special missions (10th, 16th, 20th and 26th Divisions) and the transport units (4th and 13th Divisions).

- All serial numbers beginning with 5 are reserved for 'Theater-subordinated' units such as reconnaissance, SAR, survey, and transport including the SAR- and transport brigades.

- All serial numbers beginning with 6 and 7 are for tactical assets such as fighters and ground-attack brigades or units assigned for trials and test.

- All serial numbers starting with 8 are reserved for the PLA Naval Aviation.

- All numbers starting with 9 are reserved for PLA Army Aviation Brigades.

In 2017 the newly established Theater Command transport and SAR brigades introduced another system starting with the number 5 followed by a number denoting the TC in protocol order, as here on a Z-8KA from the NTC. (Hakunamatata via PDF)

In line with the latest reorganisations, the training units reintroduced a four-digit system as seen here on this JL-8 assigned to the Xi'an Flight College, 2nd TR. (Top.81 Forum)

Similar to the training units, the Airborne Forces still use a four-digit system based on the old system reserved for the independent air regiments. (Top.81 Forum)

Individual unit markings

In 2016 the PLAAF introduced low-visibility markings for both the serial numbers and the PLAAF emblem for the first time, as demonstrated by this operational J-20A. (CDF)

One unique and interesting novelty concerning PLAAF aircraft markings is the introduction of individual unit badges and markings. During the 1990s the PLAAF briefly allowed a few divisions to show their colourful unit emblems but these were only on aircraft assigned to the then FTTC wearing the unit badge or the two premier-fighter divisions – namely the 1st and 2nd – on a few J-7E, J-8B, J-11 and Su-27SKs. Regrettably these were deleted soon afterwards.

However, in 2011 the PLAAF established the Golden Helmet air-to-air combat competition and surprisingly the winning team or team of the winning pilot is always allowed to introduce some sort of special markings which usually mirror a certain element of that unit's former regimental badge. First to use these markings were the former 44th AD, 131st AR with an eagle's head, followed by the former 33rd AD, 98th AR with a winged 33, the former 18th AD, 54th AR with a chevron framing an eagle's head and several more. Regrettably again, most of these markings were deleted once these units were re-equipped with a new type as in both first mentioned units but if still in use, these markings are shown in the relevant ORBAT chapter together with the current brigades. Fortunately, since mid-2017 regular brigades have gradually introduced such markings in order to improve the *esprit de corps* of the units.

In *Modern Chinese Warplanes* (2012), all known unit patches of individual divisions were depicted. This is now a far more complex exercise: even though the various brigades also have individual patches, very few are known.

The first unit to be permitted to wear individual unit markings was the 44th Air Division, 131st Air Regiment flying J-10As, which re-introduced the famous eagle's head. Interestingly, the red part of the head is formed by the Chinese characters '131'. (Top.81 Forum)

AIR FORCE AIRCRAFT, HELICOPTERS AND UAVS

CURRENT OPERATIONAL TYPES

This chapter is devoted to the aircraft fulfilling combat and combat support roles operated solely by the PLAAF and not Naval Aviation (for which, see *Modern Chinese Warplanes: Naval Aviation – Aircraft and Units*, Harpia Publishing, 2018). The emphasis is on types currently in service. In order to provide a better understanding of their designations, it is necessary to understand the basics of Chinese nomenclature for locally manufactured aircraft. Although this system has been adapted several times, its basic principles are similar to those in use within the US, albeit influenced by some characteristically Chinese traits.

In contrast to *Naval Aviation – Aircraft and Units*, where the description of the individual aerial assets was not as comprehensive as in the original volume and all aircraft and helicopters in service were mentioned only with comparative tables listed, in this book a detailed description of all systems – including for completeness those unique in Naval Aviation service – will be given. Due to the complexity and sheer number of systems in PLAAF service, this chapter is structured in a similar way to the original issue from 2012. However, in order to concentrate on the PLAAF's current ORBAT and particularly the most recent developments, the description of the individual aerial assets will be concentrated only on those types and variants which are still in service in mid-2018. It is less comprehensively structured as that in 2012.

Fighters

Chengdu (CAC) Jian-7 (J-7)

(ASCC 'Fishbed' and 'Fishcan')

Regardless of more modern types currently under development or in serial production it should not be forgotten that by numbers, this enigmatic fighter still forms the backbone of the PLAAF's fighter fleet. The Chengdu J-7 is still operational in five fighter versions and one trainer version and if one includes the latest JL-9 as a member of the family this type is still being manufactured and introduced into service.

Of the fighter versions the oldest currently operational type is the J-7B, which was based on the original J-7I variant built in the 1970s. It is recognisable by its original

The J-7L – this is an aircraft from the former 14th Division, 42nd AR – is the final iteration of the J-7, produced by upgrade of J-7Es to J-7G standard. (Top.81 Forum)

delta wing which it inherited from the MiG-21F-13 and it features a new rear-hinged two-part canopy which replaces the original forward-hinged and unreliable canopy of the Soviet design. This allows the replacement of the equally unreliable pilot-escape system by a more modern ejection seat. Even today the J-7B and its upgraded J-7BH or J-7H sub-versions (which are capable of using the PL-8 AAM) and the J-7K (featuring four wing pylons) are operational in several regiments and/or brigades, most of them related to the second line or to training. Next in line of the older types is the J-7D all-weather interceptor, a fighter with only mixed success which was initiated in order to develop an improved all-weather, day- and night-fighter version based on the Soviet-designed MiG-21MF.

However, after a protracted and long-delayed development during the 1980s and a limited production of only a few J-7Cs the final Guizhou-produced J-7D entered service in the late 1990s but with little success and only were 32 produced. The last J-7Ds were seen in service in late 2016 but their current status is unconfirmed, however, they are most likely to have been retired. After both second-generation models proved unsuccessful in the mid-1980s CAC once again used the trusted J-7B airframe to develop an affordable highly agile fighter, the J-7E (with the revised ASCC reporting name 'Fishcan'). Its most important modification was a new double-delta wing designed by the Northwest Polytechnic University. Together with a more powerful WP-13F turbojet all these changes led to much improved aerodynamic performance and better manoeuvrability particularly at lower altitudes. Overall, from 1993 a total of 263 machines

The most numerous J-7 variant in PLAAF service is still the J-7B, which originates in the 1970s. It seems as if it will carry on for some years in some training units.
(CJDBY)

were delivered to the PLAAF and Naval Aviation and several operational fighter regiments and/or brigades are equipped with this version.

The final Chinese MiG-21 fighter derivative then became the J-7G, a version that incorporated several airframe changes and avionics updates derived from the export models F-7MG/PG. The J-7G features further improved avionics including the KLJ-6E Falcon pulse-Doppler-fire-control radar for improved all-weather air-to-air-combat capabilities, and the one-piece windshield of the later F-7PGs to give the pilot a better view in close combat. First flown in June 2002, the PLAAF has received about 80 aircraft since 2003 and as a MLU several J-7Es were updated to a similar standard – most notably in their avionics – and are now designated J-7EG or J-7L. Their most prominent external differences are new dorsal and ventral UHF/VHF antennas similar to those installed on the J-10A/AS and some feature a VLOC antenna (i.e. an omni-directional radio range localiser as part of the aircraft's ILS system) on the vertical tail fin.

Production of the J-7 series finally ended in May 2013 with the F-7BGIs being delivered to the Bangladesh Air Force and in 2014 the last operational J-7s were retired from Naval Aviation service.

The future

For many years the J-7 was not only numerically the most important type in the PLAAF inventory but it was also exported to several other countries. Although there are still a few hundred left in service it is now certain that the story of the J-7 in PLAAF service is

The J-8 was China's first true indigenously developed interceptor to attain operational service. This is a rare example of one of the last original J-8IIs assigned to the 78th Brigade. These are in the process of replacement by upgraded J-8Hs and J-8Fs.
(Top.81 Forum)

slowly coming to an end. Currently, there are still some brigades flying the J-7B, J-7E/L and the J-7G but it seems as if the final J-7Ds have already gone and most J-7B brigades in service will either convert to J-10 or simply be disbanded. It was long expected that the last J-7B would not survive into 2018 with active fighter brigades but retirement seems to have been postponed and it is possible this variant will remain in service for some years to come, together with the JJ-7As assigned to training units.

Shenyang (SAC) Jian-8 (J-8)

(ASCC 'Finback')

Similar to its ancestor the J-7, the original J-8 also had a very long development history. Its origins stem from the requirement for a new long-range and high-altitude fighter during the early 1960s when it was decided to develop a new fighter based on the proven J-7 technology and manufacturing techniques. Concept studies began in 1964 at Shenyang but the type entered limited service only in the mid-1980s. By now all these original J-8s and J-8Is – in contrast to the sometimes even older J-7B – have been retired.

The second and now major operational version is based on the so-called J-8II and J-8B. Although it made its maiden flight in 1984, serial production did not begin until 1992. However, once again the biggest weakness of the otherwise regarded 'giant leap' for the Chinese aviation engineers was its reliance on a very dated design and this original second-generation J-8II is currently operational in only very limited service.

The next step towards the current operational version was, ironically, initiated during the brief Sino-US honeymoon in the mid-1980s which ended with the riots on

This upgraded J-8DFH, in service with 2nd Division, 3rd AR, wears the rarely seen falcon badge on its tail. This unit is reportedly under conversion to the J-16 as the 3rd Brigade.
(CMA)

Tiananmen Square and the following arms embargo. As a result, China managed to receive some help from Russia and Israel to create the J-8H and J-8F and finally the J-8G and JZ-8F during the mid to late 1990s. Both fighter versions differ noticeably with regard to their avionics suites – particularly their radar and the associated missiles these radars support.

As soon as it became clear that the multi-role J-8C would not reach serial production, the SAC worked hard to develop additional operational versions, each one essentially optimised for a specific role.

The J-8H (K/JJ8H) is based on the J-8D interceptor. The J-8H project started in 1995 as an MLU for the former J-8B/D. Its major improvements include the new Type 1491 (KLJ-1) pulse-Doppler radar with a look-down/shoot-down mode and a range of 80km (50 miles) compatible with the PL-11. Other features include INS/GPS, HOTAS, an integrated ECM suite and a modified wing with two wing fences on each side of the wing to achieve better handling. This version is powered by two slightly improved WP-13B turbojets. Following a maiden flight in December 1998 it was certified in 1999 and was introduced by the PLAAF in 2002. Additional J-8Hs were converted from existing J-8Ds and J-8Bs into the J-8BH or J-8DH. In recent years most of the remaining J-8Hs were upgraded with a new dorsal datalink antenna which is probably related to a new BM/KZ900 ELINT pod, and also new VLOC antennas on the vertical tail fin and it seems as if, as a further MLU, several have been upgraded to the J-8F standard, which also

A pristine J-8DF assigned to the 3rd Brigade jettisons its braking chute after landing following a training mission. (Top.81 Forum)

The sole dedicated reconnaissance type in PLAAF service is the JZ-8F. This example is assigned to the 93rd Brigade.
(Top.81 Forum)

introduced the PL-12 AAM. Development of the J-8F (K/JJ8F) as the final interceptor variant began in 1997. Originally the J-8F was based on the failed J-8C and it features a more modern multifunctional Type 1492 X-band pulse-Doppler radar with a look-down/shoot-down capability (75km/46 miles look-up, 45km/28 miles look-down for a $3m^2$ target) and the ability to track-while-scan 10 targets and engage two simultaneously. Other improvements besides the twin wing fences are a modified cockpit with a HUD and two small MFDs, a 573A1 INS, an IFR probe and, most significantly, more powerful WP-13BII turbojets. It was initially planned that these would be replaced by the WP-14A turbojets with increased thrust, however due to reliability issues, they appear to have been replaced. Following its maiden flight in 2000 and after testing was completed in 2002 the type entered service with the PLAAF in 2003 and with the Naval Aviation a little later. Outwardly, the J-8H and the J-8F are almost indistinguishable. One of the main identifiers compared with the older models is the new black radome with its six stripes, and the additional pair of wing fences. Some examples also feature a new datalink antenna behind the cockpit. The main difference between the two variants concerns their armament (see below). In addition to around 30 newly manufactured aircraft, Shenyang have also converted a number of older models to the standard known as the J-8DF, and the naval equivalent designated J-8FH. Following this MLU, they are in fact no longer distinguishable.

The final variant of the J-8 is the dedicated tactical reconnaissance version JZ-8F or JC-8F (re-designated in August 2012). Following repeated reports concerning a reconnaissance version this was finally unveiled in 2007. In contrast to its predecessor, the original J-8R, which carried an external camera pod, this version features a conformal camera compartment underneath the cockpit. Since a total of at least three different kinds of configuration are known, and which are replacing the original twin 23mm gun compartment, it is assumed that this fairing has a modular design, able to contain multiple cameras in different channels and angles. The optical camera is probably based on the KA-112A, while other options are an IIR camera or a synthetic aperture radar (SAR). Additionally, it features a dorsal datalink antenna in front of the vertical tail fin and several examples were upgraded with VLOC antennas on the vertical tail fin. However, it is unclear if the aircraft is capable of transmitting digital images in real time. Although rarely seen, it is known that the JZ-8F is also able to use various ELINT, SIGINT and SAR pods, which externally are similar to the KZ900, for missions at night or under poor weather conditions and the reconnaissance version uses the IFR probe as a standard fitting. This type entered service with both the PLAAF and Naval Aviation in 2006.

A defence-suppression version known as the J-8G, and based on the J-8H, was unveiled in 2009. It can be identified by the semi-spherical ESM antenna under the forward fuselage, allegedly including a device to guide the YJ-91/Kh-31P anti-radar missile. However, it seems as if the J-8G was in fact more of a test project and all current operational examples are either J-8Hs or J-8Fs.

The future

By 2012 it was assumed that the J-8 had reached the end of its development potential and that the PLAAF and Naval Aviation would soon retire it and replace it with the J-11B or even the J-16. Apparently, this process was slowed down so that even by mid-2018 small numbers of older J-8BH/DH airframes plus a handful of J-8IIs were operated alongside the latest J-8Fs and JZ-8Fs. It can, nevertheless, be assumed that the career of this important type is slowly coming to an end.

Chengdu (CAC) Jian-10 (J-10) Vigorous Dragon

(ASCC 'Firebird')

The Chengdu J-10, China's first modern, single-engined fourth-generation multi-role fighter has been under development since the early 1980s. It was originally envisaged

A J-10A assigned to the 26th Brigade taxies back after a training missing during the Sino-Russian Aviadardts exercise in August 2018. (Daniele Faccioli)

that it would replace the obsolete J-7 fighter and Q-5 attack aircraft as the PLAAF's standard multi-role fighter but was delayed for several reasons, in particular by engine issues. Aerodynamically, as a tailless-delta-canard-design, it was based on the experience gained by the Chengdu Aircraft Corporation (CAC) and No. 611 Institute on the J-9, which had been cancelled in 1980. Also, there had been undeniable secret contacts between CAC and the Israeli IAI on the cancelled Lavi most of all related to FBW development and FCS integration. Consequently, much has been written about the J-10's history, its long and protracted development based on several earlier concepts and its controversially discussed relationship to the IAI Lavi. Since some observers still assume the J-10 to be a phoenix-like resurrected Lavi, it is possible that this ancestry has led to the ASCC codename 'Firebird', which was heard for the first time in 2014. Regardless of all these discussions the J-10 became not only China's most modern multi-role fighter within the PLAAF, but also the most recent successful indigenous product of the Chinese aviation industries after many frustrating years of failed projects. It must, however, be noted that, regardless all claims, the J-10 is not a Lavi clone and nor is it a copy, since it is much larger, heavier and uses a Russian AL-31FN turbofan engine, a modified AL-31F with relocated gearboxes. This also powers the Su-27/J-11 series of fighters and provides a significant easing of maintenance and logistics.

Development of the J-10 was conceived in 1984 and a design designated as the J-9VI or J-9B, with double-delta wing and canards, as well as a chin-mounted intake which was frozen in 1986 formed the basis for the future J-10. The subsequent development proceeded very slowly, with most available funding invested in the J-8II and, during its 18-year development, the design went through at least one major redesign from the J-9B through an initial Lavi-like air-superiority fighter design to the latest multi-role fighter. The situation changed only during the early 1990s, when a decision was taken to construct a full-size wooden mock-up in order to convince politicians of the validity of the J-10. This was completed in 1991 – still with the original WP-15 and later WS-10 engine – and a prototype was set to fly in 1996. However, problems with the development of the planned WP-15 engine almost led to the cancellation of the entire project, and the type finally matured only once Moscow agreed to sell AL-31 engines to the PRC and the design had to be rethought in order to accommodate the Russian engine. The first prototype was finished in June 1997 after a 15-month delay and made its first flight in March 1998. Flight testing was completed by the end of 2003 and serial production had started even earlier that year, enabling the J-10 to enter service within the PLAAF in June 2004. Following its service entry the J-10 was produced in three batches before being replaced by a slightly improved version J-10A, which features a modified cockpit, a WL-9 radio compass antenna dish behind the canopy and, reportedly, an uprated Type 1473G fire control radar. All older J-10s were upgraded in the meantime and overall production continued in four more batches up to late 2014. The most recent images indicate that several J-10As have been upgraded with a new dorsal UHF/VHF antenna as well as VLOC antennas on top of the vertical tail fin.

As is typical for a modern fighter, the J-10 features advanced avionics including a glass cockpit with a wide-angle HUD, two monochrome MFDs and one colour MFD. It uses HMS, HOTAS, GPS/INS, air data computer, ARW9101A RWR, a Type 634 digital quadruplex FBW, digital fuel management system, mission management system, ARINC429 databus, and a detachable IFR probe. At the heart of its avionics is a new Type 1473 pulse-Doppler fire-control radar and most examples use the PL-8 short-

A J-10B once assigned to the 2nd Fighter Division, 5th AR during the Zhuhai Airshow in 2016; this unit is now the 5th Brigade. Noteworthy are the PL-12 dual missile launch rail adapters.
(Top.81 Forum)

range IR guided AAM and the PL-11 semi-active guided medium-range AAM, although this was soon replaced by the active radar guided PL-12 medium-range AAM. In order to increase the total number of PL-12s the J-10 can carry, a new twin-rail missile launch pylon was developed, which is currently being superseded by the PL-10 and PL-15. A typical load for CAS missions would be two LS-500J 500kg (1,102lb) LGBs or the newer GB1/TG500 together, since 2015-16, with a K/JDC01A laser designator pod and a K/RKL700A ECM pod. For the SEAD role there are usually two YJ-91 ARMs, one CM802AKG guidance pod and the K/RKL700A ECM pod is carried on one each of the hardpoints on either side of the intake. In mid-2018 it was confirmed that the J-10C can also carry the heavy KD-88, the PLAAF's standard missile for precision strike.

Besides the J-10A and the naval AH, there is also a tandem-seat trainer designated J-10AS (K/JJ10S/JJ-10) or, in naval use, J-10ASH. The J-10 twin-seater features a prominent bubble canopy which can be opened as a single piece and a large dorsal spine to accommodate the electronics displaced by the rear cockpit. This type is not only an advanced trainer, but also a fully combat-capable type, even if it is not used as the often-rumoured dedicated attack or EW/'Wild Weasel' anti-radiation aircraft. Following two prototypes which first flew in 2003 the J-10S was certified in late 2005 and it entered service with PLAAF soon after. The slightly modified J-10AHs are all from Batch 06. Reportedly, they differ only in their enhanced corrosion protection modifi-

To date, the most advanced J-10 variant is the J-10C. This aircraft is from the 5th Brigade, but has censored serial numbers. It carries a KD-88 training round. (Fan Yishu via chinamil.com)

cations and are allegedly able – although this is still unseen – to carry YJ-83K AShMs. Possibly this is could be an MLU option for the future. Altogether more than 224 J-10A single-seaters and about 48 twin-seater J-10ASs were built in the period 2002-14 in seven production blocks. These included the Naval Aviation J-10AH and its equivalent twin-seater J-10ASH as well as in late 2009 a few dedicated aerobatic examples called J-10AY and J-10SY which are flown by the PLAAF's Aerobatic Demonstration Team 'Ba Yi' (1st August).

The next variant to follow the J-10A was the significantly improved J-10B, which was unveiled in December 2008, and which in 2016 would be superseded by the J-10C. The most noticeable differences are a characteristic fixed diverter-less supersonic inlet (DSI/bump), which not only reduces weight but also lessens RCS, a new indigenous IRST/LR dome in front of the canopy, a further-improved glass cockpit with a wide-angle holographic HUD similar to the one fitted to the J-11B, three large colour MFDs and an HMDS. In addition, the tip of the vertical tail fin was redesigned as were the two vertical fins, featuring a long compartment housing communication and ECM antennas and, finally, a rear-facing MAWS sensor was added underneath the parachute boom.

The most striking feature, however, is a new radome, which hoses an X-band passive electronically scanned array (PESA) radar developed by the No. 607 Institute (capable of tracking 10, and engaging four targets simultaneously). The radar is said to be the first PESA type ever developed for a Chinese fighter.

The most likely reasons for the delays were problems related to the new avionics system which comprised a new fire-control radar and, once again, the engine. Although an even more eagerly awaited WS-10B-powered fifth prototype was flown in July 2011, serial production was delayed until mid-2013 and when the first production airframe was seen, it was again powered by a Russian engine; the J-10Bs are using

AL-31FN series 3 engines, which have increased thrust and are more reliable than the earlier AL-31FN.

It seems as if initially the J-10B was to be equipped with an AESA developed by the No. 14 Institute but this radar was not ready in time and so the PLAAF decided to introduce that type in a slightly downgraded version on the J-10B, which was produced as the Batch 01. One J-10C was first seen at CAC in late December 2013 numbered '2-01' although it was renumbered '1051' in December 2014. This version allegedly features further enhanced avionics including the AESA radar and more composite materials. Externally, both the J-10B and J-10C are almost identical except for additional blade antennas (datalink) on top of the fuselage and under the front section as well as ahead of the vertical tail fin and the wings' leading edges. The most prominent difference between a J-10B and J-10C is that the J-10C lacks the MAWS on the base of the tail fin and has VLOC antennas installed on top of the vertical tail. As a replacement, new RWR and MAWs have finally been added on all J-10Cs delivered since mid-2017; these are located on top of the tail and on the sides of the intake.

A total of five J-10B and one J-10C prototype had been built up to 2010 – the latest of them powered by the WS-10B – with serial production starting in early 2013 and ending in May 2015. The J-10B was built in one batch (Batch 01) of about 56 aircraft and since 2015 all Batch 02 aircraft have been called J-10C. Production of the J-10C is continuing and must by now be well within Batch 03 or even 04, however, due to increased security levels, no current information is available.

The future

Currently, a mystery surrounds the date when the indigenous WS-10 will be ready. This engine which has been under development since the 1990s at the Shenyang-based AVIC Aviation Engine Institute (or No. 606 Institute) and the Shenyang Liming Aero-Engine Group was planned for use on both the J-10 and J-11. Although almost cancelled in the early 2000s, a revised design known as the WS-10A Taihang was completed in November 2005 and since 2009 has been in service on all Chinese-built J-11 and J-16 variants. The WS-10 produces around 132kN thrust and provides the PLAAF with a useful alternative should relations with Moscow worsen. However, even though this more than eagerly awaited turbofan had already been tested on the one J-10A and the fifth J-10B prototype number 1035 in July 2011 plus two additional Batch 01 J-10Bs, it is deemed still not ready for the J-10. Consequently, all J-10B/Cs produced so far are still powered by the trusted Russian AL-31FN Series 3 and it is currently unclear if the WS-10B will be introduced in later batches. However, in late 2017 reports appeared which concerned additional J-10B prototypes (such as number 1031) being re-equipped with the WS-10 and this also included the converted prototype number 1034 as a thrust vector control testbed. This type was first flown on 25 December 2017 and is related to the development of a TVC nozzle for the J-20's WS-15. The latest rumours suggest that CAC/No. 611 Institute is already developing another upgraded variant, so far designated J-10D, which is said to feature CFTs, specialised for precision strike and SEAD, the WS-10IPE (14t class) turbofan and further RCS reduction measures. It may be a twin-seater.

As such the future seems bright for the J-10 in PLAAF service and interestingly, with production nearing 450 this year, in 2014 a small piece of the J-10's history was revealed: a grainy image was published showing a Chinese delegation including the J-10's 'father' Song Wencong standing in front of a Lavi prototype.

Only a few J-10s are currently powered by the indigenous WS-10A for test work. Even more remarkable is this dedicated TVC testbed featuring a LOAN nozzle which has been flying since December 2017. (SDF)

A J-11A assigned to the 41st Brigade photographed during a combat patrol mission over the South/East China Sea. It is armed with R-73 and R-27ET2 missiles.
(FYJS Forum)

Sukhoi Su-27/Shenyang (SAC) Jian-11 (J-11)

(ASCC 'Flanker')

For many years the PLAAF lacked a truly modern fourth-generation high-end fighter and, following the cancellation of the Peace Pearl treaty in mid-1989, the then Soviet Union was quick to step in and replace the West as China's number one arms resource. The first negotiations about a possible arms package were initiated in 1990 and these were followed by high-level demonstrations of both the MiG-29 and the Su-27 during March 1991 in Beijing. After intensive evaluations of both types, China decided to order 20 single-seat Su-27SKs and six Su-27UBK two-seat conversion trainers in the summer of the same year.

As is commonly known, the Su-27SK was chosen to become not only the PLAAF's first modern air-superiority fighter but also a really heavy fighter capable of countering the US F-14 and F-15. A contract signed in 1991 was complemented by two additional orders in 1996 and 2000. In all, the first batch included 26 aircraft (20 SKs and 6 UBKs) with deliveries starting in 1992, the second batch comprised 24 aircraft (16 Su-27SKs and 8 Su-27UBKs) starting in 1996 and the third and final batch was of 28 Su-27UBKs with deliveries beginning in 2000. This agreement included a USD1.2 billion contract to licence-build 200 Su-27SKs under the Chinese designation of J-11 by the Shenyang Aircraft Corporation (SAC).

Negotiated in 1995, the contract included a 'domestic use only' clause, thereby strictly forbidding future exports. Also, the local manufacture specifically excluded the AL-31F turbofan, for which Russia denied China a production licence. It was agreed to begin with the Chinese assembly of Russian-supplied kits and later to continue production using an increasing proportion of indigenous components, finally leading to full production in China. The first two J-11s were completed in December 1998, but these – and most of the initial production block – reportedly suffered severe deficiencies in terms of manufacturing quality and required a complete rework by Russian technicians. Licenced production was a steep learning curve, but by late 2002 the planned production rate had been achieved. It was confirmed that around 48 aircraft had been assembled by that time, and by 2003 the projected production rate of 15 to 20 per annum had been attained.

Altogether a total of about 95 kits had been delivered from KnAAPO by 2004 but once again avionics fit became the main reason for a dispute between the Chinese manufacturer and the Russian designer. China made repeated demands for Russia to upgrade the J-11 with improved avionics and weapons systems, mainly because they concluded that the N001 pulse-Doppler radar delivered by the Russians was now dated. It seems that Russia initially denied this request, or at least demanded additional negotiations regarding costs. As a result, in late 2000 SAC announced that not all of the 200 licence-built Su-27s would necessarily be built from kits, prompting speculation that production might shift to the upgraded multi-role version after about the 100th J-11. Following additional negotiations, the original radar was replaced by, or upgraded to, N001V standard which provides improved target tracking performance. This version of the radar was further succeeded by the N001VE, which is capable of engaging two targets simultaneously and employing both the extended-range SARH R-27ER1 AAM and the active radar-homing R-77 AAM. It now seems that China was pursuing a twin-track approach by requesting and partially funding a Russian upgrade programme – sometimes referred to as the Su-27SKM – while at the same time developing an 'indigenous' variant, the J-11A (K/JJ11A), which began production in 2000. A number of older aircraft have been upgraded to a similar standard, which includes a cockpit with new, digital electronics, flight information system and at least two MFDs. This variant had its maiden flight in December 1999 and, including the updating of older aircraft up to the end of 2006, a total of 105 J-11 and J-11As had been produced and modernised before production switched to the J-11B. Currently, only the PLAAF is flying the J-11A.

The J-11B (K/JJ11B) however, was a more ambitious programme, since besides indigenous improved avionics – most significantly a new Chinese multi-mode pulse-Doppler radar – the standard AL-31F engine was to be replaced by the Shenyang-Lim-

A J-11BS from the 89th Brigade deploys its huge dorsal airbrake after an aerial display during the PLAAF Aviation Open Day in September 2018.
(Longshi via PDF)

ing WS-10A 'Taihang'. It also features a modified structure with a weight reduction reported to be of 700kg (1,543lb), due to the intensive use of composite materials. The first J-11 to be powered by the WS-10 turbofan (approximately 122.58kN/27,557lb initially) was tested on a J-11WS engine testbed in June 2002. The first true J-11B prototype powered by two WS-10s flew in 2004 and altogether three prototypes (numbers 523 to 525) were tested at the CFTE, each with slightly varying configurations related to specific tests of subsystems. This led to a dispute between the Chinese manufacturer and the original Russian designers. The dispute was taken up in the more-or-less informed media about how legal this decision was if the resulting J-11B was an unlicensed copy or simply an indigenous update containing Chinese avionics and other systems already paid for but without using the original Russian parts. However, as so often before, development was delayed by technical issues due particularly to engine reliability problems and even though the J-11B's existence had been acknowledged by the Chinese government in May 2007, the first production batch needed once again to use standard AL-31F engines. Since late 2009 these problems have been solved and from the second production batch onward all J-11Bs have used the WS-10A. A land-based 'naval' J-11B variant exists which is similar to the Naval Aviation's J-10AH, and which is designated J-11BH. Reportedly the final latest batches (presumably 06 and 07) are equipped with FADEC and a further-improved version of the WS-10A, producing a higher thrust of about 129.45kN.

With regard to avionics the J-11B features a Chinese multifunction pulse-Dopple radar known as Type 1493 with a search range of up to 150km (93 miles), the ability to track six to eight targets and engage four simultaneously – a system compatible with Chinese AAMs. There is an ARINC429 data bus, an indigenous IRST/LR as well as an updated glass cockpit featuring five MFDs and a new wide-angle holographic HUD. In addition, a Chinese ECM system was installed internally, thus negating the need for the typical Russian Gardeniya wingtip ECM pods and enabling the wingtip pylons to carry AAMs. These were altered in shape and modified to a deeper version in order to carry the PL-8 with its wider fins. Allegedly, the J-11B features a new indigenous digital FBW system which initially suffered some reliability problems. Finally, the J-11B features a modern UV-band missile approach warning system (MAWS) with two sensors

The second unit to introduce individual aircraft markings was the former 33rd Division, 98th AR with an eagle's head symbolising a winged '33', as seen on this Su-27UBK and J-11A. This unit is now the 98th Brigade flying the J-16. (Top.81 Forum)

installed on both sides of the tail sting to provide coverage for the rear hemisphere. The standard AAMs are the PL-8 short-range IR-guided AAM, the active radar guided PL-12 medium-range AAM and, since 2015, the latest PL-15. Since 2017 the newest PL-X ultra-long range AAM with a range of up to 300km (186 miles), has been seen, at least on test. For air-to-ground strikes, the J-11B can carry a wide range of 90mm (3.54in) unguided rocket pods or 250kg (551lb) low-drag general purpose bombs but precision strike munition has not been confirmed. Since June 2016 it has been suggested that some have been upgraded with an indigenous datalink system similar to the American Link 16.

In parallel to the J-11B – and somewhat surprisingly since the licence agreement did not include the trainer – SAC managed to reverse-engineer the Su-27UBK twin-seater version updated to the same J-11B standard and known as the J-11BS (K/JJ11BS) and the J-11BSH. The first prototype, which was first flown in late 2007, features all the latest avionics and the latest engines. Another example appeared in 2009, but allegedly one J-11BS prototype (number 532) crashed in 2009, reportedly due to reliability issues of the new indigenous digital FBW system. However, that type was certified in May 2010 and although this version was at first expected to act as a fighter bomber this turned out to be inaccurate and the dedicated attack version is the J-16. Operationally, the J-11BS is not only assigned to J-11B units but has also replaced some of the older Su-27UBKs in J-11A units.

The future
The next step beyond the J-11B is a programme labelled J-11D. This version is currently under development at the No. 601 Institute/SAC while the J-11C is being by-passed, and it features major internal and external changes including a reshaped radome hiding an AESA radar allegedly developed by the No. 14 Institute, an improved digital FBW systems reportedly derived from the J-16 and a refined structure using increased weight-reducing composite materials in the wing and tail sections. The wing also appears to feature an additional hardpoint enabling the carriage of up to eight PL-12s or PL-15s as well as re-profiled wing-tip pylons for the PL-10. Other changes include an IFR probe installed on the port side of the windshield, while the IRST/LR is offset to the starboard side in a similar way to that of the J-16. The J-11D uses uprated WS-10IPE engines with an alleged maximum thrust of 14 tons. The first J-11D prototype (number D1101) made its maiden flight on 29 April 2015 and there are now said to be four prototypes under test at the CFTE and at Dingxin. However, the current situation is unclear, particularly following the delivery of the Su-35s and there are rumours that this programme has lost priority. There are also reports that the PLAAF wants to evaluate the Su-35 in operational service to decide if and how to continue. Another option could be that the J-11D will only be the pattern for an MLU programme of the current J-11B, which would include the AESA radar and IFR probe but without any structural changes. This upgrade – similar to that of the J-8D and H being upgraded to DF and HF standard – would then provide the Air Force and Naval Aviation with a capable long-range fighter until its successor – the J-20A – enters service.

J-11D prototype 01 clearly showing the most obvious changes compared with all other Chinese J-11 variants – the canted radome featuring a new AESA radar, the IFR probe and the offset IRST dome. (FYJS Forum)

Chengdu (CAC) Jian-20 (J-20) 'Mighty Dragon'

(ASCC 'Firefang')

The CAC J-20 is not only the PLAAF's premier fighter but also the pride of China's aviation industries. Consequently, much has been written about it and analysts, enthusiasts and spotters have been eagerly following this project's progress with great interest since its sudden appearance in late 2010. What at first was accepted with great scepticism resulted finally in a most successful project that lead to service entry in late 2016 as the world's third operational stealth fighter (after the F-22A and F-35B).

The project leading to the J-20, as it was first disclosed by the US Office of Naval Intelligence (ONI) in 1997 was originally known in the West as the XXJ or J-XX but officially in China it is dubbed the 'project 718'. It was then known that both of China's major fighter aircraft manufacturers – the Shenyang Aircraft Industry Co. (SAC) and Chengdu Aircraft Industry Co. (CAC) – had been working in competition on advanced fighter designs for some time. Understandabley, few facts became known and contradicting news items were often posted, but it seems as if the No. 601 Institute at Shenyang proposed a relatively conventional concept featuring a 'tri-plane' design with canard, widely canted tail fins and horizontal tails while the No. 611 Institute at Chengdu was researching a more radical tailless-delta-canard design with two V-shaped tails and lateral DSI intakes. This was the chosen design and, surprisingly, in November 2009 the PLAAF Deputy Commander General, He Weirong stated that China's fourth-generation fighter would fly 'soon' with a projected IOC between 2017 and 2019. (A Chinese fourth-generation type is equivalent to fifth-generation in the West).

The maiden flight was successfully made on 11 January 2011 and the first two aircraft numbered '2001' and '2002' were rated as technology demonstrators for a so-called concept/demonstration phase. Prototype '2002' appeared in May 2012 and many observers were concerned about a longer break before the next aircraft appeared. In the meantime CAC began weapon integration tests showing both the new PL-10 IIR guided short-range AAM on a unique retractable side missile launch rail and the new PL-15 long-range AAM. Before the next aircraft was unveiled, prototype '2002' was re-numbered to '2004' and '2001' received a new light grey coat of paint and both were transferred to the CFTE at Xi'an-Yanliang for further testing. After a long delay, the third prototype appeared in late 2013 as a radically modified aircraft featuring 'major improvements'. In quick succession, to the end of 2014, four more prototypes (numbers '2011' to '2015') made their maiden flights and many observers were now expecting a quick delivery of additional aircraft of this initial standard to soon join the flight-testing. However, this proved not to be and the seventh aircraft numbered '2016' was seen for the first time on 11 September 2015 with what was to be the final prototype, numbered '2017', making its maiden flight on 24 November 2015, after which all were delivered to the CFTE and FTTC for further testing. One reason for the delays could be due to the CAC's request for a more powerful version of the Saturn AL-31FN Series 3 that now powers the all the current aircraft.

In contrast to the first two demonstrators these '201x series' prototypes feature several refinements including stealth coating, a redesigned intake, re-profiled vertical stabilisers and smaller underwing actuators to further reduce RCS. They also have two enlarged tail booms – once again reshaped from 2013 – which house additional EW/ECM and self-defence systems such as two pairs of chaff and flare dispensers to

In parallel with the introduction of low-visibility markings, the PLAAF began experimenting with different J-20 colour schemes. Of the six LRIP J-20As confirmed to date, two each have the standard grey scheme, two wear this splinter scheme and two are painted in the so-called 'Raptor scheme' with larger blocks of colour. (rabbit via CDF)

The public unveiling of the J-20A was at the Zhuhai Airshow in 2016, when two of these fighters made a brief appearance in front of the crowd. Unfortunately both fighters were not numbered to hide their true assignment. (Minghua Wei)

protect the rear hemisphere of the aircraft. Other modifications include a reshaped canopy which was once again reprofiled with '2017', and which features an additional inner frame as well as an EOTS similar to the one used by the F-35, and a retractable IFR probe hidden underneath a cover on the starboard side of the cockpit. The cockpit itself is a modern glass type with either three large colour LCDs or one 24 x 9in (610 x 229mm) touchscreen panoramic cockpit display (PCD) allegedly controlled by voice command similar to that of the F-35. There is also a smaller LCD between the pilot's legs and a prominent, often greenish-glowing, wide-angle holographic HUD. A true HMD was finally confirmed to be operational in May 2018. Also, the J-20 is controlled by a side-stick and throttle and the combination of AESA radar, EODAS, IRST, advanced cockpit design and HMD further improves the pilot's situation awareness by providing the highest degree of 'information fusion'. Overall, the J-20 has a very smooth surface with no protruding pitot tubes or inlets, suggesting a FADS has been installed, and the latest aircraft show two flush hexagonal-shaped side fuel-air heat-exchangers in order to reduce RCS. Noticeable also are two small diamond-shaped windows on both sides of the nose, which allegedly house EO sensors. To provide a full 360 degree coverage these are complemented by two additional sensors underneath the rear fuselage as well as by two more located forward and aft of the cockpit, suggesting a distributed situational awareness system similar to the EODAS fitted to the F-35. Its overall weight is made up of 20 per cent titanium or titanium-alloys and 29 per cent composite materials. The remaining 51 per cent is of aluminium and different steel-alloys. The J-20 carries its weapons in internal bays; there is one large central weapons bay for up to four PL-15 medium/long-range AAMs and two smaller lateral

The current J-20As operational with both the 172nd and 176th Brigades are all thought to be LRIP aircraft like this as yet unnumbered aircraft.
(Top.81 Forum)

weapons bays behind the intakes accommodating one PL-10 short-range AAM each. Additional stores can be carried underneath four hardpoints for further AAMs or even fuel tanks. The pylons are said to be able to be jettisoned in order to quickly turn into the aircraft into the stealth mode. A gun is not yet confirmed although some observers assume a compartment for this was reserved underneath a top panel on the port side of the fuselage next to the canard wing.

Little is known of other systems: the new Type 1475 (KLJ-5) fire-control radar is expected to be an AESA type under development by the No. 14 Institute. In parallel to the J-20 prototypes from '2013' onwards it is currently in test on board a heavily modified Tu-204C airliner acting as a dedicated radar testbed at the CFTE. This is similar to the US Boeing 757 testbed for the F-22 and features the J-20's front section and radome but also several other EW and communications systems in a huge 'canard' or wing-like structure mounted on top of the airliner's fuselage. Additionally, a similarly converted Y-8C assists the testing of other sub-systems. Complementing the main radar, the J-20A has probably two additional side-looking arrays installed underneath the elongated hexagonal dielectric fairings on each side of the nose, which provide better situation awareness and extended missile guidance during a dogfight. Otherwise, the final two prototypes '2016' and '2017' differ in certain details and represent the final standard prior to LRIP. Their engines no longer have their previously typical silverish metallic sheen and the nozzles now appear in a dark dull charcoal and the side weapons bays seem to be different as well.

The biggest uncertainty remains its engine, since the final WS-15 Emei seems still to lag behind the fighter's schedule and, following the most reliable reports, will not appear until 2022-24. Consequently, much has been written about this issue and the exact type of engine remains unclear. However, based on the most reliable reports, it is believed that the demonstrators were powered originally by AL-31FNs, while the prototypes from '2011' onwards were using AL-31FN Series 3s, similar to the J-10B/Cs which were produced in parallel. But since the J-20 is referred to as a fifth-generation fighter and the known thrust values given for the FN Series 3 are deemed too low for such an aircraft, this would seem to result in a still underpowered fighter. The J-20 is therefore suspected of lacking true supercurise capability at least until the planned WS-15 turbofan enters service. However, the PLAAF is well-satisfied with the current powerplant so that for the LRIP models and current serial J-20As, the situation might be somewhat different. Following some reports, the J-20 might use a specially tailored version of the Salut AL-31 based on the AL-31FM2 which stems from a secret joint development between China and Salut similar to the once secret original AL-31FN agreement for the J-10. For the thrust figures – if these follow the data quoted for regular FM2s – about 142kN can be assumed, or roughly the same as the 117S used in the Su-35. This theory, however, is not confirmed but in the author's opinion, the Sino-Russian military cooperation stands on more solid ground than most expect, more than most are willing to accept and even more are able to officially confirm. Whatever the truth, it seems to be confirmed that even if limited in its capabilities due to its lacking the definitive engine, the PLAAF is eager to bring this type into service as soon as possible to exploit and explore operational tactics and procedures for this new fighter.

Again, surprisingly close to the prototypes '2016' and '2017', the LRIP (or batch 00) aircraft numbered '2101' appeared in late December 2015 making its first flight on 18 January 2016. In parallel, several J-20s were undergoing weapons testing at Dingxin, high-altitude testing at Daocheng-Yading in mid-September 2016 and several J-20s actually took part in several exercises against other PLAAF assets. Reportedly – albeit with no fanfare or public celebrations – the first LRIP J-20As were handed over to the PLAAF on 12 December 2016 at Dingxin in a new 'tactical' light grey paint with low-visibility markings but also sporting a striking, splinter and a Raptor-scheme. In February 2018 the J-20A officially entered PLAAF service in a front-line trials unit at the Cangzhou Flight Training Base.

The future

The most important issue remains its powerplant, since the definitive WS-15 is still some years off, despite there being repeated rumours suggesting that a J-20 is preparing for the integration of the WS-15. However, if the latest reports are correct, then the CAC is not testing the WS-15 but an interim engine based on the WS-10 Taihang. The first prototype (number 2021) was seen in September 2017 fitted with a variant called WS-10X (sometimes also known as WS-10B or C or IPE), which is built by Shenyang Liming. This develops a thrust of about 140 to 145kN (31,473 to 32,597lb) and has, for the first time, serrated nozzle feathers. A second prototype ('2022') flew for the first time in January 2018 and one additional 201x-prototype has been fitted with TVC nozzles since early April 2018. This may suggest further proof of fighter engines 'Made in China' and their testing achieves another milestone on China's way to full independence from its Russian counterpart.

Regrettably, since early 2017 the much stricter internet security rules have reduced the almost constant flow of news and images from Chengdu Huangtianba, with the result that the current status of both the J-10 and J-20 programmes is no longer clear. Neither do we know how many aircraft have been built in recent months or how many

The next variant likely to enter service after the first Batch 00 J-20s will be a variant powered by the WS-10C featuring a LOAN nozzle.
(FYJS Forum)

have been delivered. If the tests are successful, it can be expected that the next batch of serial J-20s – possibly the J-20B – will use the WS-10X, until later in the decade when the definitive WS-15-powered variant appears. In addition, it has been reported that the WS-15 completed ground testing in August 2015, with a thrust of about 160kN (35,969lb) attained, and it was said to be ready to test on the IL-76LL platform 'soon'. However, other reports suggest that development was delayed and therefore the WS-10 was chosen as an interim measure. It seems likely that the WS-15 will not be ready for the J-20 until 2022-24 at the earliest and consequently it makes sense to replace the Russian engine for the next batch of 01 aircraft.

It remains to be seen if the J-20 will also be introduced into Naval Aviation service, if there is the possibility of a carrier variant, and what other applications the PLAAF will find for that type. Since early 2018 there have been reports of a tandem-seat version being under development.

KnAAPO/Sukhoi Su-35

(ASCC 'Flanker')

Following the acquisition of the Su-27SK/UBK and Su-30MKK/MK2 and the many Chinese developments of them, the latest purchase of a Russian 'Flanker' family member seems surprising to many observers. But after what felt like endless rounds of negotiations, repeated rumours which no-one was believing and heated discussions among the observing community, a deal involving the delivery of 24 Russian Sukhoi Su-35s was signed between China and Russia on 19 November 2015. The first aircraft were delivered to China on 25 December 2016 and by mid-2018 four batches of a total of 19 aircraft had been handed over, with the final five expected to arrive the same year.

Overall, this deal is more than mysterious. It was officially confirmed by the Chinese at a very late stage, but had been regularly announced by Russian media including Russia's state technologies corporation Rostec. It is reportedly estimated to be worth USD2 billion – or about USD83 million per unit and it makes the PLAAF the first foreign customers of these aircraft, which were previously operated only by the Russian Air Force, and it is one of the largest contracts for military aircraft signed between both countries. There are other peculiar issues: first, why should China purchase yet another new Russian variant in parallel with its own 'Flanker' developments including the J-11D and the J-20? Why only so few and what about the concerns that China might again illegally reverse-engineer the Su-35 or its engines? Possible reasons are China's interest in exploring the TVC engines and their operational use and also to use the Su-35 for dissimilar air combat training (DACT) against other PLAAF assets as was the case during the most recent Red Sword exercise. Another argument might concern urgent operational needs particularly in the disputed South China Sea area and finally for as yet unknown political reasons. In consequence, it is difficult and probably erroneous to name a single reason for the Su-35 contract. It is hardly believable that this deal was only a cover framework for the acquisition of the 117S powerplant and it would be wrong to conclude anything about the status of China's own engine programmes such as the uprated WS-10 and particularly the WS-15 planned for the J-20. It does seem, however, that this contract brought benefits for both sides: Russia finally had a first export customer for the Su-35 and the PLAAF obtained a new and potent

Two Su-35s armed with either R-77 or R-77-1 AAMs providing long-range escort for an H-6K cruise missile carrier patrolling over the West Pacific and over the South China Sea. Officially, four batches of 19 aircraft had been handed over by mid-2018 and the final five were expected to be delivered by the end of 2018.
(Shao Jing via chinamil.com)

fighter assigned to an operational unit, which brought about new capabilities. It provided China with an insight into the latest Russian avionics and propulsion systems; it gave the PLAAF for the first time the opportunity to explore TVC and to develop operational tactics and, finally, it deepened the Sino-Russian political cooperation.

Technically, the Chinese Su-35s – although not the Su-35K or Su-35SK – are equipped in a similar way to their Russian armed forces counterparts. At the heart of the avionics system is the Irbis-E PESA radar, the same MAWS and LWR sensors installed in the forward fuselage, a quadruplex digital FBW system, 117S turbofan engines with TVC nozzles and they are even equipped with the latest L-265M10-2 Khibiny-M ECM pods on the wingtips. The only visible differences are a deleted middle navigational antenna mounted to the rear edge of the right vertical tail fin and the use of a hitherto unknown EO pod which could be either some sort of camera or a training round of an also unknown type of EO guided bomb or air-to-surface missile. As for weapons, the Chinese Su-35s were seen only armed with R-77-1 and R-73/74 AAMs. The latest news suggests that the PLAAF wants to explore the operational use of this first batch against the J-11D after which it might sign for a larger purchase.

Fighter aircraft currently in use with the PLAAF

Type	Role	Service entry	Avionics/radar	Main weapons	No. (est.)
J-7B/BH	Fighter	1980/1986	Type 222	PL-5IIE, PL-8	288/96
J-7C/D	Fighter	1988/1995	JL-7A	PL-5IIE, PL-8	Retired?
J-7E/L	Fighter	1993/201x	Type 226/ KLJ-6E	PL-5IIE, PL-8	120-144/24
J-7G	Fighter	2004	KLJ-6E	PL-5IIE, PL-8	72-82
J-8II	Interceptor	1992?	Type 208	PL-8, PL-11	Retired?
J-8B/D	Interceptor	1996	Type 208B	PL-8, PL-11	12, modified
J-8H/BH/DH	Interceptor	2002	Type 1491	PL-8, PL-11	some 48

J-8F/HF	Interceptor	2003	Type 1492	PL-8, PL-12	some 48
JZ-8F	Tactical reconnaissance	2006	Type 1492	PL-8, PL-12, EW pods	24–30
J-10/J-10A	Multirole	2006	Type 1473	PL-8, PL-12, LS-500J	up to 240
J-10AS	Multirole/trainer	2010	Type 1473	PL-8, PL-12	
J-10B	Multirole	2014	Unknown PESA	PL-8, PL-12, YJ-91, LS-500J, EW pods	56
J-10C	Multirole	2016	Unknown AESA	PL-10, PL-15, YJ-91, KD-88, LS-500J, EW-pods	some 80, in poduction
J-11/J-11A	Air superiority	2010	Type 1493	PL-8, PL-12	some 100
J-11B	Air superiority	2010	Type 1493	PL-8, PL-12	some 120-180?
J-11BS	Air superiority/ trainer	2010	Type 1493	PL-8, PL-12	some 90
J-20	Air superiority/ multirole(?)	2016	Type 1475	PL-10, PL-15	
Su-27SK	Air superiority	1992	N001E; later N001P	R-73E, R-77E	some 20; retired?
Su-27UBK	Air superiority/ trainer	1996	N00E1 later N001VE	R-73E, R-77E	16
Su-35	Multirole	2016	Irbis-E PESA	R-74E, R-77-1	24 until December 2018

Fighter-bombers and bombers

Nanchang (now HAIG) Qiang-5 (Q-5)

(ASCC 'Fantan')

The real 'dinosaur' in the PLAAF ORBAT is the Nanchang Q-5 close air support fighter. Its history began in 1955, when the PLAAF issued a requirement for a more 'modern' replacement for their venerable Il-10 attack aircraft. Although initiated at the Shenyang Aircraft Design Department, development was soon transferred to the Nanchang Aircraft Factory (NAF), where it evolved into the well-known Q-5. The design was finalised in February 1959 but due to a turbulent political situation development was delayed until 1963, leading to a maiden flight in June 1965. Overall, considerable development work was carried out in the following decades and the Q-5 became the standard CAS type within the PLAAF and Naval Aviation in several different versions. Of these the so-called Q-5IA model entered service during the mid-1980s and was further improved to the Q-5B to include an improved gun- or bomb-sight, a laser-rangefinder and RWRs; sometimes the designation Q-5C is also used. Succeeding the Q-5B and C, the next model became the Q-5D, which is based on two failed Westernised ver-

The Q-5L was the last and most capable variant, recognisable on account of the small fairing underneath the nose which houses an ALR-1 laser rangefinder/laser spot tracker. It is also able to carry laser-guided bombs guided by a K/PZS-01H targeting pod.
(Acer31 via CDF)

sions which were initiated before the Tiananmen riots and the resulting embargo. It was developed in the 1990s and entered service late in the decade. The main difference to its predecessors were enhanced avionics with improved fire-control and navigation systems including HUD, GPS/INS navigation, RWR and a ballistic computer. Interestingly, it replaced the majority of the obsolete earlier Q-5 versions in PLAAF Ground Attack Divisions, but ultimately it was retired earlier than the oldest Q-5IA/B/C versions, which are due to be retired by the end of 2018. Naval Aviation has already replaced all its Q-5s with the JH-7.

Always regarded the biggest weakness of this type in a modern warfare environment was the lack of any precise strike capability and although a few aircraft were identified as Q-5E and F to explore such an option, it was not until the late 1990s when the first Q-5L appeared. In contrast to all other versions, the Q-5L – which was built by upgrading Q-5B/C models – features a small fairing underneath the nose which houses an ALR-1 laser rangefinder/laser spot tracker as well as a strengthened fuselage pylon for a K/PZS-01H targeting pod. The main weapon is the LS-500J 500kg (1,102lb) LGB under the wings and some have also been seen carrying a huge semi-conformal fuel tank under the belly to achieve a longer range, which could be even further extended by two large drop tanks. The Q-5L has also been seen carrying the Russian-made KAB-500Kr-U captive training system for the KAB-500Kr and KAB-1500Kr LGBs. The latest addition to its weapons arsenal seems to be a new 250kg (551lb) LGB – possibly called GB-3 or TG-250 – featuring a proportional navigation seeker.

The final version was a twin-seat variant developed during the early 2000s to replace the obsolete JJ-6 fighter-trainers and for use also as forward air controllers. This type has a completely redesigned forward fuselage with two separate side-opening canopies and an enlarged tail fin to improve stability. All of these are rebuilt from retired Q-5D single-seaters.

Most Q-5Ds have been converted as Q-5J twin-seaters, to serve as trainers or as fast forward air controllers like this aircraft from the former 28th Division.
(FYJS Forum)

The future

Although the original Q-5 has been withdrawn from service, the availability of significant numbers of Q-5 airframes has made this type an attractive, economical option for further upgrades. Several of the earlier Q-5Bs and Q-5Cs are being rebuilt to the latest Q-5L standard, while a number of retired Q-5Ds have been converted as Q-5J twin-seaters. Since the Q-5J designation is unique – conventionally it would have been a QJ-5 or Q-5S – some Chinese sources suggest that the latter variant might be deployed as a fast forward air controller. In 2017 a hitherto unknown version, Q-5N, was mentioned for the first time, allegedly featuring a navigational upgrade similar to the latest J-7L.

Following rumours from late 2016, the PLAAF had decided to retire all its last active Q-5s 'soon'. At first it was impossible to deduce how credible these reports were. In early 2017 the PLAAF still operated seven Q-5 regiments and/or brigades with about 200 aircraft. Finally, in mid-June the PLAAF did indeed abolish its first Q-5 unit and by March 2018 most, if not all, Q-5s had been retired. This may be in the case of only one regiment, but it is likely that all units but one have been either disbanded or converted to J-7E. If this trend is correct, then the Q-5 will be gone before the end of 2018.

Xi'an (XAC) Hongzhaji-6 (H-6)

(ASCC 'Badger')

The history of the H-6 began with the agreement of the Soviet government in early 1956 to help in building up a Chinese medium bomber fleet. This decision was largely influenced by politics and an agreement was signed in September 1957 to create an assembly line in Harbin for the Tu-16 together with the provision of all necessary technical documentation. Following the delivery of the first two Tu-16s in January 1959 as pattern aircraft, the first Tu-16 in kit form was delivered to Harbin in May. Assembly began almost immediately with the help of specialists from Kazan and additional qualified workers assigned from Shenyang. This phase of the H-6 development continued in a relatively problem-free manner and after the delivery of the Soviet-made parts had begun, the first 'Chinese-built' H-6 – basically a Tu-16T/Tu-16A hybrid – took off in September 1959. The first two aircraft went through the typical series of factory acceptance trials and were later delivered to the PLAAF/Naval Aviation for operational testing. During these evaluations at least one of the aircraft was modified with an air-conditioned bomb-bay to carry nuclear weapons. The final testimony of the success of these modifications was the dropping of a live nuclear bomb at the Lop Nor test site in May 1965. As was typical for the time, political influence delayed this promising and important programme after 1961: the reason being the decision to change the manufacturing site from Harbin to Xi'an. While Xi'an should concentrate on the new H-6, it was decided that Harbin – where a major Il-28 facility was already located – should start to reverse-engineer the Il-28 as the H-5. Moving the complete H-6 production from Harbin to Xi'an was no easy task and the biggest challenge was to relocate the already installed machinery and tools and this was not completed until 1964. To make things even worse, during this transfer a large amount of technical descriptions and manufacturing documentation was lost. To the detriment of the already cooling Sino-Soviet relationship each side was quick to blame the other. There followed a great effort to retrieve the missing data to which end the two Soviet-delivered Tu-16s were dismantled and studied in detail to reverse-engineer them. This lasted far longer than anticipated and manufacturing preparations in Xi'an were begun only in 1964 with completion of the first static-test airframe in October 1966. Finally, nearly nine years after the first Chinese H-6 made its first flight in China, the indigenous aircraft made its maiden flight on 24 December 1968. In order to distinguish these new Xi'an-built airframes from the first two from Harbin they were given the designation H-6A, and these were delivered only to the PLAAF. In parallel to the standard bomber version, the H-6 was used for test purposes including engine tests and, in a similar way to the Soviet Tu-16LL known as Type-226, it was modified to a drone carrier and specialised sub-types for reconnaissance were also used.

Development of the basis for the first variants remaining in service began during the mid to late 1970s, when the H-6D specialised anti-shipping missile carrier was devel-

The H-6H – this example is assigned to the 29th AR – is the oldest bomber in active PLAAF service and is likely to be withdrawn from use before long. This type is recognisable by its characteristic fairing underneath the rear fuselage, which is thought to contain the datalink antenna for the KD-63 missile.
(Top.81 Forum)

oped for the Naval Aviation. This version could be distinguished by two large wing-pylons carrying the YJ-6 ASM and an enlarged bulbous flat-bottomed new Type-245 attack radar. Production of this version started in about 1980 and although already superseded by more advanced sub-types, it formed the basis of nearly all then-current operational versions and the final remaining H-6Ds were converted to H-6DU tankers. The PLAAF, however, followed a slightly different path and modernised its H-6As with updated navigation systems and uprated ECM and ESM equipment during the late 1980s. This version became operational as the nuclear-capable H-6E and the conventional H-6F. Most, if not all, original H-6A and E/Fs are now withdrawn from operational use and today only a handful are reportedly flying alongside the updated H-6M in one mixed regiment, although in mid-2015 one training brigade received a few for the training of bomber crews.

After all attempts to obtain a more advanced Russian type or to develop an indigenous long-range bomber had failed, the PLA was left only with the option to once again modify its venerable H-6s. Following the service introduction of the H-6D during the 1980s, this version formed the basis of all further modified versions operated as stand-off missile carriers.

The first modernised version is the H-6H for the PLAAF and it is operated as a dedicated stand-off missile carrier carrying two KD-63 LACMs. It is recognisable by a new tear-drop radome behind the bomb bay and the lack of all defensive armament. Development of this version began in January 1995 and the first prototype was rolled out in April 1998. Following its maiden flight in December 1998 the H-6H became operational in 2002. Since its service introduction it has replaced all former subtypes and was long seen to be the final service variant and would extend the life of this 50-year-old design well into the 21st century. However, during the last few years the H-6H was gradually replaced by new-build H-6Ks and it remains to be seen if they will be modified or reassigned to other tasks. A few remaining aircraft received a moderate upgrade to their avionics to bring them into line with the H-6G – some say they are now H-6HG – this includes RWR antennas installed on the vertical tail and chaff and flare dispensers on the rear side fuselage. Some also have VLOC antennas installed on the tail fin and the original TV-guided KD-63 has been replaced by the upgraded IIR-guided KD-63B with improved accuracy and all-weather performance.

The final H-6D-derived version operated by the PLAAF is the H-6M. This subtype evolved as a mid-life update version converted from retired H-6E/F bombers and it became operational in 2007. Similar to the naval H-6G, it includes the large chin-mounted radar and two pairs of underwing pylons as well as a modernised cockpit including MFDs and a much improved self-protection suite including RWR antennas on the vertical tail fin tip. It has MAWS sensors mounted forward on the nose and prominent chaff and flare dispensers along the side of rear fuselage – something together with a new dorsal UHF/VHF antenna which the other two operational versions have received in only in recent years during depot maintenance. Its biggest difference compared with the H-6H, however, is its ability to carry two KD-20 ALCMs and therefore the H-6M is believed to act as a low-cost, stopgap solution until being fully replaced by the more advanced H-6K. Additionally, the latest KG600/700 ECM pods can be carried on smaller outer wing pylons and there is another small pylon underneath the rear fuselage behind the bomb bay door which could carry an AKK-802K datalink pod associated with the KD-63 ALCM. This missile was not confirmed to be included

First introduced in 2007 as a stopgap solution until the H-6K entered service, the H-6M evolved from refurbished H-6E/F bombers modified to carry two KD-20 ALCMs.
(F.KSCAN via Top.81 Forum)

in this type's weapon's arsenal, but images from early 2015 assume it is so. In the mid-term it is expected that all H-6Hs and H-6Ms are to be replaced by the H-6K.

This most recent variant – and in fact a completely refurbished model – was wholly unexpected until its prototype was unveiled as a dedicated cruise missile carrier in mid-December 2006. After being almost unchanged for decades, the H-6K represents a complete redesign featuring most significantly a new airliner-style non-glazed nose with a huge radar and slightly wider air intakes. These were necessary, since the H-6K is powered by two Russian D-30KP-2 turbofans, also used by the Il-76MD transport aircraft. This not only eases maintenance because of a common engine but also boosts the bomber's performance since the new turbofans are far more efficient than the former WP-8/AM-3 turbojets, which results in an extended range, a higher cruising speed and/or an increased weapons load. Besides these major external changes, the use of composite materials reportedly reduces weight and its cockpit has been completely redesigned and features six colour MFDs and ejection seats for the three crew. The H-6K is equipped with a chin-mounted EO turret containing FLIR, CCD TV camera

and a laser designator for night/poor weather missions, a SATCOM antenna above the rear fuselage and a datalink antenna below the rear fuselage. A redesigned tail cone contains defensive electronics including tail MAWS sensors (RKG963A) and RWRs and an APU, while other avionics identified around the fuselage include MAWS and RWR. The usual chin-mounted surface search radar has been replaced by a new large ground scanning PESA radar developed by the No. 38 Institute allegedly featuring SAR and TF/TA capabilities. However, the most important change is a new weapons system for a reported maximum of six KD-20 long-range cruise missiles. Some sources assume that a seventh missile can be carried in the main weapons bay, but this is not confirmed. Besides the KD-20, the H-6K can carry also the older KD-63 and even a mix of both missiles has been noted. The H-6K is also able to carry additional ECM pods such as the KG600/700 or 800 underneath additional pylons on the outer wings. Little is known of its background, other than that the development of this new version was officially begun in 2003 and that two prototypes were built. The maiden flight took place on 5 January 2007 and flight testing presumably continued until April 2011, when the H-6K entered service. Additionally, it was long rumoured to be funded by XAC without the PLAAF's support. The reason for this were concerns by the PLA since that would make the future of the whole programme entirely dependent on Russia's willingness to supply the D-30KP engines. Overall, 55 D-30KP-2 engines were imported from Russia between 2009 and 2011 and this engine is allegedly being reverse-engineered as the WS-18 by the Chengdu Engine Corporation. However, nothing is officially known on that type and an additional (perhaps around 184) D-30KP-2 engines have been reportedly ordered but a solution for their domestic production remains uncertain.

In parallel to the different bomber versions, the PLAAF studied several other roles for that trusted airframe. The most mysterious development is the adoption of the H-6 airframe for an EW/ELINT version called HD-6 which has been the subject of rumours since 1990, especially since a model of a very much modified H-6 was seen featuring a solid nose and a large canoe-shaped fairing under the fuselage similar to that of the Tu-154M/D Type III. Although this type is so far unconfirmed, it reportedly existed in three different subtypes with the latest version known as HD-6III. Also, some former H-6Ds have been seen carrying huge ECM/EW pods following their retirement from the naval attack role.

The most recent and probably the most surprising model, however, is the H-6LV – meaning Launch Vehicle – or spaceplane carrier. Reportedly, since 2000, the China Aerospace Science & Industry Corporation (CASIC) has been studying an air-launched, all-solid-propellant, three-stage space launch vehicle (SLV) similar in appearance to the US X-37 which can place a 50kg (110lb) payload into earth orbit. During the 2006 Zhuhai Air Show a mock-up of this air-launched SLV was finally shown and in December 2007 a photograph was released showing a former H-6A carrying a small rocket-powered SLV known as Shen Long (Divine Dragon).

The future

The future of the H-6 in PLAAF service would seem to be secure for several years to come, despite the fact that it is thought to be only a stopgap type until the new H-20 stealthy strategic bomber enters service later in the next decade. Since 2016 there have been rumours concerning several new variants. The first is said to be a dedicated AShM carrier variant currently under development for the PLAN Naval Aviation and

Since 2015 a dedicated H-6N AhBM carrier has reportedly been under development, equipped with a nose-mounted IFR probe and able to carry a single large ballistic missile underneath the fuselage. Yet another variant unveiled in July 2018 might carry two smaller hypersonic AShBMs. (APFSDS)

When the original licenced H-6 flew in China for the first time in 1959, few would have expected the type to be in production nearly 60 years later. This H-6K represents a dramatic evolution from the Tu-16. (Shao Jing via chinamil.com)

aimed to replace the H-6G. This type, sometimes referred to as H-6KH, H-6N or even H-6J, reportedly flew as a prototype in 2014 but was confirmed only in August 2017. In contrast to the standard H-6K it is equipped with a prominent nose-mounted inflight refuelling probe enabling it to be supported by the newly acquired Il-78 tanker and is able to carry up to 6 YJ-12 AShMs. The most important new capability, however, is a single large AShBM – similar to, but larger than, the recently unveiled Russian Kh-47M2 Kinzhal – carried externally underneath the fuselage in a semi-recessed fashion. So far, this missile has not been confirmed but reports vary between a variant of the DF-21D – sometimes called DF-21E, with a range of up to 1,500km (932 miles), a DF-15/-16-based design or a variant of the DF-12 (M20), which could be used against US aircraft carriers. Other potential roles could be a conversion for the long-range ASW role or tanker. Latest reports suggest that the H-6KH and H-6N are in fact two different variants with the H-6KH now being the H-6J for the Naval Aviation and the H-6N being an IFR-capable H-6K.

Xi'an (XAC) Jianhong-7 (JH-7) 'Flying Leopard'

(ASCC 'Flounder')

Similar to its predecessor, the Xian JH-7 was also born out of a requirement formulated after a military conflict, when, in 1974, around the so-called Xisha Islands the PLAN and the South Vietnamese Navy were engaged. Here, the requirement for a modern replacement for the Q-5 and H-5 received additional urgency. Following a RFP issued in 1975, development was hampered by internal rivalry between the PLAAF and the Naval Aviation , which both favoured different main performance requirements as well as airframe configurations. In the end, XAC was authorised to take on the development since this proposal offered a compromise between technological risk and operational performance, which would result in an early in-service date. Again, delayed by technical issues and increasingly by the political situation following the Cultural Revolution, the design was frozen in late 1983. It was decided that the aircraft would be powered by the Rolls Royce Spey 202 and later with a licensed version called WS-9, a decision which probably saved the JH-7 from being cancelled. Its maiden flight took place in December 1988 and, after an intensive flight test programme, the JH-7 entered service in 1994. In the meantime, ongoing issues – particularly with the JH-7's main sensor, the Type 232H Eagle Eye multifunction fire-control radar – led to the PLAAF's withdrawal from the programme. This move was unsurprising since the PLAAF was now able to order the Su-27SK and, in 1999, it acquired the specialised Su-30MKK fighter-bomber version too. As a consequence, the JH-7 became the first dedicated maritime attack aircraft for PLA Naval Aviation.

Following a few so-called Block 01 aircraft (approximately 18) – possibly regarded as pre-serial numbers only – in the early 2000s the engine issue could be resolved by purchasing another 90 used Spey 202 engines and a second production run of Block 02 aircraft followed. On this improved model, the unreliable and less-than-powerful Type 232H radar was replaced by the new JL-10A multi-mode pulse-Doppler radar with enhanced air-to-air and air-to-ground modes. Besides its original Type 232H Eagle Eye multi-role radar, the JH-7 features a triplex digital-analogue autopilot, an 8145 air-data computer, WG-5A radio altimeter, 210 Doppler navigational system and HZX-1B 'stabilising' system. Additionally, the comprehensive EW suite includes RW1045 RWR, 960-2 noise-jamming system, and 914-4G passive jamming system. As such, this JH-7 Block 02 (of about 20 aircraft) can be seen as a step between the original JH-7 and the projected definitive serial version to be called the JH-7A. By now all Block 01 aircraft have been upgraded accordingly.

When the problems with the WS-9 Qinling were finally solved in 2007, the way was open for the definitive production version, the JH-7A, which had been under development since the mid-1990s and also the PLAAF came back to the 'Flounder' due to its precision strike capability. The reasons for that seem to be that although impressive in its performance, the Su-30MKK was simply too expensive to be deployed in the required large numbers and, secondly, the 'Flanker' was still not compatible with Chinese-designed missiles such as the YJ-8 series of ASMs. In contrast to the original JH-7, the JH-7A differs in some important modifications made to the airframe. The most prominent external differences are a new one-piece windshield, a re-profiled wing without the wing fences and two large fins under the rear fuselage instead of the old central fin.

Besides these, it has additional pylons under the mid-fuselage below the air-intake for the K/JDC01 targeting pod and a datalink pod for KD-88 ASM and YJ-91ARM; and it also has the ability to use PL-8 AAMs on additional wing pylons. Weight-saving measures and the use of composite materials reduced the empty weight of the aircraft by 400kg (882lb), so that the JH-7A can carry a wide range of ordnance, giving an enhanced capability to launch precision strikes using anti-radar missiles, long-range missiles and also laser-guided bombs. Other less visible changes may be even more important since the JH-7A features a new databus and INS/GPS system, a new glass cockpit with an updated HUD and a new digital, dual-redundant FBW system giving the JH-7A a true terrain-following capability. Since 2009 most JH-7As have been upgraded with a new UHF/VHF antenna behind the cockpit.

Besides being a dedicated striker, the JH-7A can also carry large EW pods similar to the US ALQ-99 jammers used by the EA-6B and EA-18G and several different types of pods have been identified which differ in their antenna shapes. Following the latest information, they are part of a frequency jamming system that consists of one receiver pod and four different transmitter pods, each covering different frequencies. Standard weapons for the Naval Aviation JH-7As are, in addition to the usual PLAAF weapons, PL-8 AAM, LS-500J LGB, KD-88/KD-88A TV/IIR guided ASM and YJ-91 ARM and, most significantly, the YJ-83K/KH AShM, and MKC-03-500 250kg (551lb) aerial mines. A standard fit is also the K/JDC01 targeting pod and a datalink pod for KD-88 ASM and YJ-91ARM, and for ECM missions it carries the typical ECM and ELINT/SIGINT pods. In contrast to the PLAAF, the Naval Aviation JH-7As have not been seen carrying the latest KG600 ECM pod or the bigger KG800 ECM pod for self-defence, which can provide the electronic protection for the attacking formation.

The future

The final variant JH-7B is something of a mystery. Originally expected to be a completely new design featuring a stealth-optimised fuselage including a diamond-shaped forward fuselage and DSI intakes, the JH-7B turned out to be merely a limited MLU package containing only improvements in the areas of avionics, the flight control system and the weapons arsenal. Known to have been under development at XAC's design

The JH-7A – this example is assigned to the 15th Brigade and participated in the Sino-Russian Aviadarts exercise in August 2018 – is China's first true indigenously developed fighter-bomber.
(Daniele Faccioli)

institute No. 603 since the mid-2000s, the JH-7B retained the external appearance of earlier models. Reportedly, the main features are a new fire-control radar and mission computer as well as a new full-authority digital FBW and a retractable IFR probe under the port side of the cockpit. It is possible that additional composite materials are used to replace some parts for greater strength and to further reduce the weight and, finally, some reports assume the engine to be an improved WS-9B delivering a slightly higher thrust. Overall, the situation surrounding this type is somewhat vague since there are few images available. Furthermore, no news has appeared since its maiden flight in 2011 despite its being reported to have entered service in late 2015. Two prototypes are known and are numbered '821' and '822' at the CFTE in Xi'an-Yanliang. However, since 2015 nothing more has been heard, so that in consequence it seems to have been abandoned or, at best, will form the basis of an MLU package to keep the JH-7A up-to-date until a true successor is available. Also, this type seems to face tough competition from the J-16 multi-role fighter.

Previous reports suggested that Xi'an had begun development of a dedicated electronic warfare variant of the JH-7 for PLAAF service in 1998. This was understood to be capable of carrying three different pods equipped with ECM systems or with ELINT/SIGINT equipment. In fact, this new variant emerged as a mere update to the standard JH-7A and not a dedicated version.

Most likely a failed project or downgraded to an MLU project at best, the JH-7B was once planned as the next development of the JH-7A. Now, however, it seems as if the PLAAF has selected the J-16 instead.
(FYJS Forum)

KNAAPO/Sukhoi Su-30MKK/Su-30MK2 and Shenyang (SAC) Jian-16 (J-16)

(ASCC 'Flanker')

One of the driving reasons behind the efforts to develop an indigenous 'Flanker' version was the missing multi-role performance. Additionally, the JH-7 with its long and troubled development during the late 1980s was always in real danger of failing to fulfil its once promising aspirations in becoming a true multi-role fighter-bomber suitable for both branches. As such, and since it became apparent to the PLAAF that the ordered Su-27 designed as an air superiority fighter would be able to perform only secondary attack missions with 'dumb' munitions and not with the desired precision strike capabilities, the PLAAF was looking for an alternative design.

A solution was found with the opportunity to purchase a version of the multi-role Su-30MK and the negotiations, which began in 1996, resulted in an initial order of 38 Su-30MKKs in late 1999 for USD1.85 billion. This PLAAF-tailored version reveals some differences compared with the Indian Su-30MKI, especially the lack of canards and AL-31FP TVC engines as well as the MKI's N-011M phased array radar. In contrast, it is equipped with an uprated NIIP N001VE fire-control radar capable of guiding the R-77 AAM and is enhanced by additional air-to-ground modes. Typical Russian-made weapons are LG- and TV-guided ASMs such as the Kh-29T or Kh-59ME, Kh-31P ARMs and TV-guided bombs including the KAB-500KR or KAB-1500KR. In addition, there are Russian Sorbtsiya ECM pods and the APK-9 datalink pod. The Su-30MKK and Su-30MK2 feature the typical strengthened landing gears which are recognisable by two tyres on the front gear, considerably increased maximum take-off weights and maximum weapons loads of eight tons. The vertical tail fins are taller and of larger area: made of carbon fibre composites, they include additional fuel tanks with a capacity of 280 litres

(62 imperial gallons). The Chinese Su-30s feature Su-35-style tail fins with squared tips. One reason for purchasing a slightly downgraded variant in comparison to India's decision, was its faster delivery schedule. Even before this contract was completed by the end of 2001, it was supplemented by a second order for an additional 38 aircraft in July 2001 worth USD1.5 billion. For the Naval Aviation a third contract for 24 specialised naval-attack versions known as Su-30MK2s was signed in January 2003. These feature the upgraded N001VEP radar especially tailored to fire the Kh-31A and Kh-59MK AShMs and all Su-30MK2s had been delivered by August 2004. They are usually operated as long-range interceptors carrying R-73 and R-77 AAMs over the East China Sea. In addition, they and the MKK are employed to fly long-range escort missions for H-6K bombers over the West Pacific Ocean as well as the South China Sea and the Sea of Japan, supported by Il-78 tankers – extending their already impressive range to 5,200km (3,231 miles) – and KJ-2000 AEWs. It is rumored that they can act as airborne command posts to direct up to 16 other aircraft via datalink to engage enemy aircraft. And finally, several Su-30MKKs are wearing two different 'tropical' camouflages in tan-brown and tan-green similar to Vietnamese Su-30MKVs and are serving as Blue Force/aggressors for dissimilar air combat training.

The Su-30MKK and MK2 represented some of the most capable fighters; however, they were still limited to Russian weapons only. Consequently, for some time there have been reports about a modernisation programme including a revised cockpit with new flight instruments and domestic MFDs and an avionics update. The cockpit has additionally been upgraded with a small display for GPS/Beidou next to the HUD. Images from 2015 have regularly shown Su-30MKKs carrying the new KG600/KL700 ECM pod and Su-30MK2s armed with the PL-12 AAM. These are clearly a hint that this programme is already under way and that more Chinese weapons including the

A J-16 sporting its new low-visibility scheme thunders away for its next mission. Development of the type took some time, but it is now in Batch 04 production. By the end of 2018, close to 100 J-16s could be delivered. (PDF)

LS-500J LGB, YJ-91 ARM and KD-88 ASM will become compatible with this type. Some rumours even go so far as to suggest that the original Russian radar has been replaced by a Type 1493 as used by the J-11B, but this has not been confirmed.

Similar to the indigenous improved fighter version J-11B and the J-11BS twin-seater based on the Russian Su-27UBK's airframe (which was never built under licence and for which a licence was at least never officially granted) it had been speculated since 2010 that a dedicated multi-role strike 'Flanker' is under development. This version was unveiled in late July 2012 and appeared indeed to be based on the Su-30MKK since it featured a retractable IFR probe on the port side of the front fuselage and which required the IRST/LR system to be moved slightly offset to the starboard side. It also featured the Su-30MKK's reinforced twin-wheel nose gear to cope with the increased weight and the taller vertical tails, albeit without the typical squared fin-caps. As with all current Chinese 'Flankers' the J-16 is powered by two WS-10A turbofans. The avionics of the the J-16 reportedly feature an ASEA fire-control radar from the No. 607 Institute behind a new grey radome without a pitot. Also, new ECM systems are installed which are compatible with all Chinese-made precision guided weapons such as the KD-88 ASM and LS-500J LGB which are still excluded from the Su-30MKK's current arsenal. Other new weapons and stores are the new PL-10 SR-AAM, carried on smaller wing-tip pylons, the PL-12 and the latest PL-15 AAMs. For air-to-ground stores the J-16 has also been seen with a huge standoff submunition dispenser (perhaps the 500kg/1,102lb TL500/GB6) which is similar to the American AGM-154 JSOW, possibly a new wind-corrected munitions dispenser (WCMD) similar to the American CBU-103/104 and a new targeting pod known as YINGS-III underneath the engine intake which is similar to the American AN/AAQ-33 Sniper pod. However, it is still unclear if the J-16 is capable of carrying even bigger weapons, such as the 1,000kg (2,205lb) TG-1000 'bunker buster' LGB. The most recent new weapon is a large ultra-long range AAM known as PL-XX for use against targets such as AEW&C aircraft and tankers.

The maiden flight of the J-16 was made in October 2011 and by early 2013 at least two prototypes were seen undergoing tests at the CFTE. Two years later in October 2013 the first pre-production models numbered '161x' were seen at SAC, reportedly preparing to enter limited service with the PLAAF but, due to unknown reasons, these aircraft entered service only in May 2015. Allegedly the reasons were radar issues and although the original radar – said to be developed by the No. 607 Institute – was tested on board the J-11B prototype number 524 in 2014 with the usual pitot removed, it seems as if from Batch 02 in mid-2016, an improved AESA radar has now been fitted. Besides the new AESA fire-control radar, the J-16 differs from the Su-30MKK by being powered by two WS-10A turbofans. Also, the J-16 is fitted with a new glass cockpit featuring a prominent single-piece panoramic touch-screen display.

The future

China has so far turned down several Russian offers for advanced radars and even additional orders for a rumoured Su-30MK3 variant, and the J-16 seems to be the main reason. Also, it was expected that the SAC would develop an indigenous quadruplex fly-by-wire system for the Su-30MKK, designed to accommodate WS-10 engines, and an entirely new, indigenous avionics system. However, given the latest developments, it can be expected that all Su-30MKKs and MK2s will be superseded by the indigenous J-16 in the future and that all Russian Su-30s will be either redistributed to second-line units or possibly retired. Additionally, following its delayed service introduction, the J-16 is now in production at SAC and it seems as if it won the contest over the JH-7B to become the next multi-role striker due to its more powerful AESA radar, its greater weapons load and its longer range, although this has not yet been confirmed. So far, all serial J-16s have been assigned to the PLAAF, but it is expected that the J-16 will also enter Naval Aviation – as the J-16H – in the near future. Following the latest images, the J-16 is the second type in PLAAF service which introduced a new medium grey colour scheme with low-visibility PLAAF insignias and serial numbers similar to those of the J-20s. Consequently, the multi-role J-16 is to become the PLAAF's Tier 1 aerial combat asset together with the J-20 for multi-role missions complemented by the J-10C and UAVs.

A J-15D prototype. This variant shares many similarities with the PLAAF's J-16D EW and SEAD variant. According to rumours, both projects could be merged into a single type known as the J-17. (meyet.net)

The final version of a Chinese 'Flanker' is a dedicated SAED/EW version of the J-16. Although based on the J-16 airframe it features several modifications including a new radar behind a shorter radome, which is said to be yet another type of AESA radar enabling integrated EW modes similar to the US AN/APG-79. Also noticeable are several EW aerials, dielectric panels and multiple antennas mounted on its fuselage, on the side of the engine intake, and both behind and underneath the cockpit. There is also a rectangular dielectric panel behind the radome as well as huge ESM/ELINT pods on the wing-tips similar to those of the AN/ALQ-218 tactical jamming receiver onboard the EA-18G. However, to enable installation of the comprehensive EW suite, the typical IRST/LR system in front of the windshield and the 30mm gun appear to have been removed. Consequently, the J-16D has to rely solely on AAMs such as the PL-10 and PL-15 for self defence or on escorting fighters. The first J-16D made its maiden flight on 18 December 2015 and reports from February 2017 suggest that a second prototype has flown. It is expected to be operated in a similar role to the US EA-18G, to escort standard J-16s. Although not yet seen, the J-16D will probably carry YJ-91 ARMs as well as the new generation of ARMs (allegedly based on the PL-15 or a version of the CM-102). The J-16D is said to be related to the J-15D, which could share similar airframe changes and electronics, albeit mated with the carrier-capable fuselage of the J-15S and by late 2017 rumours apperaed, which suggested the J-15D and J-16D projects have been merged into a single type called J-17.

The J-16D prototype seen after its maiden flight in December 2015. As a dedicated SEAD/EW variant of the standard J-16 it is recognisable thanks to a pair of large wingtip ESM/ELINT pods comparable to the EA-18G's AN/ALQ-218 tactical jamming receiver. (meyet.com)

Fighter-bombers and bombers currently in use with the PLAAF

Type	Role	Service entry	Avionics/radar	Main weapons	No. (est.)
Q-5B/L	Ground attack	1983		bombs	Retired?
Q-5L	Ground attack	Early 2000s		bombs, LGBs	Retired?
H-6H	Bomber	2002	Type 245	KD-63, bombs, EW pods	48
H-6M	Bomber	2006/07	Type 245?	KD-20, bombs, EW pods	28
H-6K	Bomber	2011	Unknown new radar	KD-20, KD-63, bombs, EW pods	some 78; i/p
JH-7A	Strike	2004	JL-10A	PL-5IIE, PL-8, KD-88, YJ-91, bombs, LGBs	132–144
J-16	Multirole/strike	2014	Unknown AESA	PL-10, PL-15, PL-XX	some 80–90; i/p
Su-30MKK	Multirole	2004	N001VEP	R-73E, R-77E, PL-12?, Kh-29, Kh-59, KG700	74

Future combat types

J-XY – future medium fifth-generation fighter

For some time there has been a heated discussion about whether the next generation carrier-based stealth fighter will be a version of the SAC FC-31 currently under development and whether the PLAAF is also interested in the type. For the naval requirement, the alternative could be either a navalised version of the J-20 or even a new design whose configuration is not yet decided. However, with the success of the J-20 during Red Sword 2016 and, bearing in mind that such a navalised J-20 might be difficult to operate from the smaller Type 001 and 002 vessels, there still might be a chance for the FC-31. Similar considerations for the PLAAF are for a smaller complement to the large, heavy and expensive J-20 and also an enlarged carrier-based variant based on the FC-31 design. This has been discussed as the J-21 or even J-25 from SAC powered by two WS-19s or a new – eventually single-engined – type from CAC using one WS-15. In consequence, it remains to be seen upon what type the Naval Aviation and the PLAAF decides. In addition to the two projects that have already reached 'hardware status', there are more concepts in the making or under consideration including a new strategic bomber and several smaller bomber and fighter-bomber concepts.

Xi'an (XAC) H-XX/H-20

Although it is too early to say for sure since little is known and even less confirmed by official sources, a new long-range strategic bomber – officially known so far only as the 'strategic project' or most likely H-20 – has been under development at the No. 603 Institute and XAC since the late 1990s or early 2000s. Reportedly, the PLAAF was undecided for some time about the requirement and what type of bomber would fit this best. Consequently – and in a similar way to the alleged new tactical or regional bomber design under development at SAC – various designs were studied which ranged from supersonic configurations through conventional to quite innovative concepts. One of these is said to feature a delta wing geometry and canards while others

are more conventional and there is also a subsonic stealthy flying wing design. Following the most reliable reports, several scaled-down models have been built and test-flown so that by 2011 the chosen configuration was a four-engined flying wing design similar to the US B-2 or the B-21. Concerning other technical details, one can only speculate but the engines will most likely be modified WS-10As, without afterburners as an interim solution or probably a WS-15 derivate later. Otherwise, the new type is said to have a range dramatically longer than the current H-6K's (estimated to be more than 10,000km (6,214 miles) with a combat radius of over 5,000km (3,107 miles) and being IFR capable) while at the same time being able to carry a heavy weapons load (less than the B-2A's 23 tons but more than than that of the H-6K) for both nuclear and conventional stores. The few images published so far suggest a single centre weapons bay – although some artworks depict two bays – which will be able to carry at least six KD-20 ALCMs or any other precision strike munitions on a rotary launcher. It is expected to feature a modern avionics system built around an AESA radar with conformal antennas similar to the US AN/APQ-181 LPI radar. Additionally, the H-20 is said to feature a modern EW capability acting as a C4ISR node and to interact with other sensor platforms like UAVs, AEWs and strategic reconnaissance aircraft to share information and target data (data fusion).

Besides General Ma Xiaotian's public confirmation in 2016, a first hint of the H-X's progress was a news report from December 2015 within the AVIC group referring to a digital 3D mock-up of a 'major project' being completed, and currently the design is in the phase of detailed design-engineering. It was again noted in early 2017 that a quality control system/platform for this 3D mock-up – by now called the prototype – had been established. Also noted was the fact that 'China's next generation bomber has entered the detailed design-engineering stage'. Currently, it is assumed that construction of a prototype has begun at XAC and roll-out can be projected to 2019–20.

H-18/JH-18 and JH-XX/JH-17

Also, in parallel to the H-20 – possibly one of the smaller and finally rejected concepts – a new twin-engined flying medium or 'regional bomber' is often mentioned. This H-18 or JH-18 became known via a mysterious black model unveiled in 2016 at SAC and in early 2018 another medium-sized and medium-range semi-stealth fighter-bomber design was revealed. This type had also been rumoured for some time and has been under development at the No. 601 Institute and SAC since the 2010s. It is sometimes called JH-XX, J-17 or even JH-17 and was allegedly loosely based on the J-16, featuring a twin-seat cockpit similar to that of the Russian Su-34, but following the latest reports has evolved into this new design. So far, the only known details are a cockpit section with two seats and there are two engines. Another recent theory claims that this smaller concept was in fact the original one proposed in 2013 but rejected. However, some time later, the design was approved and has subsequently been transferred to No. 603/Institute XAC for further development into the aforementioned larger 'regional bomber'. Although this practice seems to be unique or at least unusual by Western standards, transferring projects to different institutes that are assigned more appropriate specialisation is common practice in the state-owned Chinese aerospace sector, or at least it was in the mid-2000s. And since XAC has more experience with bombers, it is not unlikely in this instance. Finally, reports emanating since late 2016 suggest that GAAC and HAIG are working on a light attack/CAS aircraft but again, no details are available and the suggestions have yet to be confirmed.

Trainers

Nanchang (now Hongdu/HAIG) CJ-6

(ASCC 'Max')

Although often mistaken for a licenced Yak-18A, the CJ-6 is a Chinese design based on its predecessor, the Nanchang CJ-5, which was in fact a licence-built Yak-18. Development began in late 1957 in order to address the shortcomings of the Yak-18A and the first flight of a CJ-6 was completed on 27 August 1958 with the aircraft powered by

Three CJ-6A trainers assigned to the PLAAF's Sky Wing aerial demonstration team.
(Top.81 Forum)

a Czech-built horizontally opposed piston engine. The engine was later replaced in the production variant by a locally manufactured version of the Soviet AI-14P (260hp) radial engine. In 1965 the HS-6 engine was upgraded to 285hp and re-designated the HS-6A. It is estimated that more than 3,000 CJ-6s were delivered to the PLAAF and Naval Aviation including some for export and it is still in production. It has long been expected that the CJ-6 would be replaced by the Hongdu/Yakovlev joint developed CJ-7 trainer, which first flew in late 2010 but, based on the latest reports, the PLAAF has ultimately cancelled this project and decided to issue a request for a proposal for a new 'next generation primary trainer' with both Hongdu and Guizhou expected to be the main contenders.

Hongdu (HAIG) JL-8

The JL-8 (K/JJL8) – which is also known as K-8 Karakorum – was a joint development initiated in 1986 by Hongdu Aviation Industrial Group and Pakistan Aeronautical Complex (PAC) for a light jet-powered primary trainer. In its original form, it was planned to feature several US-made items, most significantly, a Garrett/Allied Signal

TFE731-2A turbofan and avionics systems. However, due to the political developments following the Tiananmen Square riots in 1989, this fell through. A first prototype was built in 1989 and its first flight took place on 21 November 1990. Flight testing continued between 1991 and 1993 and the first aircraft were delivered in 1994 to Pakistan. Due to the necessary engine replacement, the indigenous JL-8 – now powered by a Ukrainian Ivchenko AI-25TLK with 16.87kN thrust – for the PLA was delayed until 1995/96. Later aircraft are using a locally manufactured WS-11 (in fact a licence-built AI-25TLK) and this version is sometimes unofficially known as the JL-11. More than 400 JL-8s are in service at the PLAAF and Naval Aviation (as the JL-8H) in the flight colleges where they have replaced the obsolete JJ-5s. They have a maximum take-off weight of 4,468kg (9,850lb), a maximum level speed of 800km/h (497mph), a maximum climb rate of 30m (98ft) per second, a maximum range of 2,140km (1,330 miles) and ceiling of 13,600m (44,619ft).

Guizhou (GAAC) JJ-7

(ASCC 'Mongol')

Besides the standard J-7 fighter versions, the PLAAF operates a dedicated twin-seater trainer version called JJ-7 which was developed by the Guizhou Aviation Aircraft Corporation (GAAC) although the first trainer variant JJ-7 was jointly developed by the Guizhou Aircraft Design Institute and the Guizhou Aircraft Company. In contrast to the much earlier MiG-21U, the new aircraft incorporated a similar new aft cockpit, a new forward-view periscope, larger twin ventral fins, fuel tanks inside the upper part of the fuselage, a modified intake, improvements in the ejection escape system, air

The JL-8 is not only used as a basic trainer, but also for weapons training as demonstrated by this example carrying a 23mm gun pod and two rocket pods.
(FYJS Forum)

A JJ-7A assigned to the 78th Brigade serving as a trainer in this J-8H/F unit. Otherwise the JJ-7A is still the most numerous jet trainer in the Flight Colleges and conversion training units. (Top.81 Forum)

conditioning, a new fuel system, red cockpit lighting, an interphone system, and a failure simulation system. The conceptual design was begun in 1979 and was completed by 1983 with trial production beginning the same year. Static testing of the airframe was finished in May 1985, fight testing began soon after and it entered service in the late 1980s. In the late 1990s, a modernised subtype of the JJ-7 was developed which was based on the export version FT-7P. Unlike the FT-7P, the JJ-7A does not feature the additional fuselage stretched by 610mm to create space for an additional fuel tank and an internal gun. However, it is able to carry a 23mm gun pod and more than 100 had been produced by March 2017. Finally, the latest member of the enigmatic 'Fishbed' family is also a development of the original J-7/JJ-7.

Guizhou (GAAC) JL-9 'Mountain Eagle'

Development of the JL-9 (K/JJL9) began in 2001 as an advanced lead-in fighter-trainer based on the earlier JJ-7/FT-7 design. Consequently, it was initially known as the JJ-7B and later JJ-9. In contrast to its predecessor, the JL-9 features a new front fuselage replacing the circular front intake by a solid nose accommodating an X-band pulse-Doppler fire-control radar with a range of 30km (19 miles), and the intakes are moved to the sides. Also redesigned is the stepped tandem cockpit section which is under a one-piece windshield giving better forward and downward views in comparison to the original JJ-7. Also new is the double delta wing without leading edge flaps and there are modern integrated avionics and cockpit (HUD plus MFD), RKL-206A RWR, ECM, 1553B databus, INS/GPS, JD-3A TACAN, WL-11 radio compass and an air data computer. However, in order to save costs and also development time, the JL-9 uses the same WP-13F(C) producing a maximum thrust of 4,400kg (9,700lb) or 6,450kg

The JL-9 has many similarities with the proposed Super-7, once planned as a major J-7 development in cooperation with Grumman.
(PDF)

(14,220lb) with reheat. Its flight control system is mechanical rather than FBW, thus limiting the improved performance of its competitor, the JL-10, which is technologically more advanced but is more expensive and was delayed in its development phase. Its normal take-off weight is 7,910kg (17,439lb), with a maximum of 9,800kg (21,605lb) and the maximum weapons load is 2,000kg (4,409lb). Its maximum speed is Mach 1.5, with a maximum level speed of 1,100km/h (684mph), the maximum load is 8G, the ceiling 16,000m (52,493ft) and the maximum climb rate is 260m (853ft) per second.

The first prototype made its maiden flight on 13 December 2003 and the prototypes were evaluated between 2004 and 2005. Following a redesign, the first revised JL-9 flew for the first time on 23 August 2006 and featured a new stability control augmentation system (CAS) and an improved cockpit environment control system. The first JL-9s were delivered in 2007 to the PLAAF for further testing and the first operational deliveries began in 2011, with the Naval Aviation receiving its first JL-9H soon after. A slightly improved variant – possibly the JL-9A – featuring new EL formation light strips on both forward fuselage and vertical tail fin for night training missions and a new VLOC navigational system has been in production since 2014 and that production continues. Following a report from September 2016, GAAC is allegedly planning to develop the next generation advanced trainer which could compete with JL-10 LIFT from Hongdu.

Hongdu (HAIG) JL-10 'Falcon'

The Hongdu JL-10 (K/JJL10) – originally, and still for export, designated L-15 – is an advanced jet trainer developed by Hongdu with technical assistance from the Yakovlev OKB. Intended as a truly modern advanced jet trainer (AJT), the JL-10 was designed

The JL-10 is almost certainly the most important future jet trainer in PLAAF service. Remarkably, it evolved from the original Russian Yak-130, as did the Italian Leonardo M-346 Master.
(PDF)

in order to support the new generation of Chinese fighters such as the J-10, J-11, J-15, J-16 and J-20. From the outset it was planned to develop two different versions: the standard JL-10/L-15A (AJT) powered by two Ukraine AI-222-25 turbofans and a dedicated lead-in fighter trainer (LIFT), powered by two AI-222-25Fs with afterburners giving this L-15B supersonic capabilities. The normal take-off weight of the L-15 is 6,500kg (14,330lb), with a maximum of 9,500kg (20,944lb), the maximum speed is Mach 0.95/1.4, the maximum climb rate is 150m (492ft) per second, g limits are +8/-3, ceiling 13,000m (42,651ft), loitering time two hours, maximum range 2,600km (1,616 miles) and the structural life is 10,000 hours.

As a new design based on the Yak-130 the JL-10 features a more modern airframe – especially in comparison to its competitor the JL-9 – with prominent leading-edge root extensions (LERX), a large vertical tail fin and an overall modern design expected to give the aircraft a high AOA of up to 30 per cent. Also, its cockpit is much more advanced since it is a fully digital glass cockpit with HUD and three colour MFDs, HOTAS control and a three-axis quadruplex digital FBW. In order to be used also for weapons training the JL-10 features four underwing and two wingtip pylons for a wide variety of stores and the option to carry a gun pod underneath the fuselage. The standard JL-10 features a small radar while the L-15B could even carry a small PESA fire-control radar.

The first L-15 AJT prototype was completed in September 2005 and its first flight was made – albeit delayed due to engine issues – on 13 March 2006, powered by two interim DV-2 turbofans and the first improved AJT prototype powered by two AI-222-25 turbofans first flew on 10 May 2008. However, the development of the LIFT version still suffered from the slow progress of the afterburner-equipped AI-222K-25F. It finally took off for the first time on 26 October 2010 and differs from the JL-10 by a stretched front section and a longer rear section and has two AI-222K-25F turbofans with afterburner nozzles. The first true domestic AJT version designated JL-10 – and which had

been rumoured since November 2012 – was finally unveiled before its first flight on 1 July 2013. Since then, several prototypes have undergone testing and from images from mid-2016 it was assumed that this type was in serial production for both services, an assumption finally confirmed in early 2017. So far, the type is in service with the PLAAF, and the Naval Aviation received its first aircraft as the JL-10H in March 2017.

For the future it seems as if the L-15 will be powered by the domestic engines. The standard JL-10 is planned for use with an indigenous turbofan without afterburning (actually an AI-222-25 copy), which was flown for the first time in May 2016. The L-15B is said to use the WS-17 Minshan turbofan with a maximum thrust of 4,700kg (10,362lb) with afterburning, developed by the Guizhou Aero Engine Research Institute and that engine was rumoured also to have been tested on a L-15 prototype. Since September 2013 there have been rumours that the JL-10 might be adopted by Naval Aviation as a carrier-based trainer possibly based on the L-15B with the more powerful engines but this has not been confirmed.

Xi'an (XAC) Y-7

(ASCC 'Coke')

The final trainer operated by both the PLAAF and Naval Aviation is a design based on the improved Y-7-100C2 passenger aircraft and designated Y-7LH or HYJ-7 (K/JYL7H) and was chosen as a simple and less expensive trainer for navigators and bombardiers of the H-6 bomber after the obsolete HJ-5s were retired. Little is known about its development, but the Y-7LH first flew in the late 1990s and is in limited service with both the PLAAF and Naval Aviation bomber regiments or flight academies. In contrast to the standard Y-7 it features a prominent, bulky observation bay with windshields attached to the starboard side of the fuselage simulating the nose of the H-6 for a single trainee.

It is equipped with the HM-1A bombing sight and DMW-1 bombing sight stabiliser and it also features the TNL-7880 composite navigation system. Uniquely, it is fitted with two long fairings or weapon bays along both sides of the lower fuselage for carrying up to 20 small practice bombs. In recent years the type has been upgraded with an enlarged fairing underneath the fuselage to train the crews of the more modern H-6H and H-6G and probably housing a new surface search radar. Also, small fairings are installed along the bottom of the fuselage, allegedly for datalink and radio compass antennas.

Images of HYJ-7 trainers are quite rare, especially ones showing the unique bulky observation bay on the starboard side of the fuselage simulating the nose of the H-6. (Top.81 Forum)

Training aircraft currently in use with the PLAAF

Type	Role	Service entry	No (est.)
CJ-6A	Basic trainer	1960s	180-190
JJ-7/JJ-7A	Lead-in fighter trainer	1989/mid-1990s	some 100
JL-8	Primary trainer	1996+	170
JL-9	Lead-in fighter trainer	2011?	some 50, i/p
JL-10	Advanced jet trainer	2017-18	some 20, i/p
H-6A	Bomber trainer	mid-1970s	12?
HYJ-7/Y-7LH	Navigator- and bombardier trainer	Late 1990s	48?

Support, transport and liaison aircraft

Nanchang/Harbin/Shijiazhuang Y-5

(ASCC 'Colt')

The Y-5 is a Chinese licensed version of the Antonov An-2. Originally, China began producing the An-2 as the Y-5 at Nanchang with several hundred having been built when production was transferred to Harbin in 1968 and later also to Shijiazhuang. Altogether, the Y-5 was built in several variants and the final version was the Y-5B-200 – sometimes also known as the Y-5C – featuring characteristic triple tipsails ('feathers') on the upper wing tips, which allegedly improved the climb rate by 20 per cent and improved L/D ratio by 15 per cent. Today still several are operational for training and liaison.

Xi'an (XAC) H-6 tanker

(ASCC 'Badger')

Besides being used as a bomber, the Xi'an H-6 is also available as a specialised tanker conversion, which exists in two different versions. The first is based on rebuilt H-6D naval bombers converted into the tanker role as H-6DUs or HU-6Ds and so still feature the standard glazed nose together with the big chin radome. In contrast to its former bomber role its weapons system and pylons were removed, but near the wing tips there are two RDC-1 hose-drum refuelling-pods similar to those fitted to RAF VC-10 and Tristar tankers. It seems that this programme ran for almost 15 years, but because of the limited numbers of aircraft available – fewer than 20 – it was first recognised in the West only when such a tanker made an aerial demonstration together with two J-8D fighters on a parade in Beijing on 1 October 1999 to celebrate the 50th anniversary of the PRC. These tankers are assigned only to the Naval Aviation and fly alongside the H-6G to support the J-8DH and DF IFR probe-equipped fighters.

The second dedicated tanker was first seen during the late 1990s and is designated HU-6 or H-6U. It differs from the older version in having a solid nose equipped with

An extraordinary image demonstrating the contrasts within the PLAAF today. On the one hand, modern equipment - here in the form of Segway personal transporters for the ground staff of an air base - and in the background a flight line of the very oldest Y-5Cs assigned to the Northern TC transport brigade
(Li Laifeng/Wang Hongkang/ Xiong Huaming via chinamil. com)

a conventional weather / navigation-radar and no chin-radome. Again, regarding its background, this first dedicated Chinese tanker – reportedly known as Project 8911 – stems from a PLAAF requirement following the clashes between Chinese and Vietnamese naval forces over the Spratlys in 1988. After failed attempts to obtain or produce under licence a foreign aerial refuelling system, the PLAAF studied several different concepts and in the end the H-6 was chosen to be fitted with two underwing RDC-1 refuelling pods enabling it to refuel two J-8Ds simultaneously. In addition, all necessary signal and illumination lights were installed for night refuelling. The H-6U was first revealed in 1988, about the same year it entered development, the first flight was made in 1990, the first in-flight refuelling took place in 1992 with a J-8D and currently about two dozen newly built H-6Us have been operational since 1996 in one regiment. During the last few years, most aircraft have received new navigational equipment and flight control systems comprising INS and TACAN, as well as gaining the same upgraded self-defence equipment including MAWS sensors, RWR antennas and chaff and flare dispensers.

An HU-6 tanker assigned to the 8th Bomber Division refuels a J-10C armed with PL-10 and PL-15 AAMs.
(Top.81 Forum)

Although the HU-6 represented a giant leap in IFR-capability compared to other modern tankers, the internal fuel capacity is somewhat limited. More seriously, following the failed Il-76/78 contract from 2005, the PLAAF still lacks a capable tanker, particularly for the Su-30MKK. As a stopgap it has managed to obtain three refurbished ex-Ukrainian Air Force Il-78 tankers in recent years, equipped with three UPAZ-1A refuelling pods and a vastly improved fuel capacity.

These are clearly an improvement, but the possession of only three tankers is insufficient for the PLAAF and Naval Aviation. As a consequence, it was long speculated that the HU-6 would receive the same engines as the H-6K or even that a dedicated H-6KU would be developed, but currently the most likely option will be a dedicated tanker version of the new Y-20 transport.

Xi'an (XAC) Y-7

(ASCC 'Coke' and 'Curl')

The Y-7 is a passenger transport developed as a licenced version of the Antonov An-24, which dates back to 1966, when XAC started a project for local production of the An-24. The first Chinese-assembled An-24T made its maiden flight on 25 December 1970 and limited production was begun in 1977 but was delayed due to technical issues and the Cultural Revolution until about 1982. Production aircraft (akin to An-24RV) were not flown until February 1984 and the majority of deliveries went to the PLAAF. These aircraft were powered by two WJ-5A-1 turboprops.

In the meantime, the Y-7 was already outdated, so that in the mid-1980s XAC initiated a project to upgrade it with Western avionics, which resulted in the Y-7-100. Besides minor changes, it is externally identifiable by its winglets. In military use this variant became the tactical transport version Y-7H (also known as Y-7H-500 in its commercial nomenclature), which was developed in the late 1980s as an unlicenced variant of the An-26 incorporating a fully pressurised fuselage, a rear cargo ramp, two improved WJ-5E turboprop engines, an auxiliary turbojet fitted to the left engine, improved cockpit avionics and other military equipment. The first Y-7H made its maiden flight in 1989 and it entered service with the PLA in the late 1990s.

One originally civil-only offspring of the Y-7 is the MA-60 (Modern Ark 60) medium transport aircraft, which is a stretched version of the civil Y-7-200A and able to carry up to 60 passengers. The aircraft made its maiden flight in 2000 and is now in PLAAF and Naval Aviation service as the Y-7G. It is used mainly to transport military personnel between the Hainan Island and the Paracel or Spratly Island in the South China Sea and as a VIP transport by the PLAAF.

Shaanxi (SAAC) Y-8

(ASCC 'Cub')

This type is based on the original Antonov An-12, is similar to the US C-130 and has become the PLA's standard tactical transport. The Y-8 is one of the few aircraft projects which survived the Sino-Russian break in 1960 and although the immediate Russian advisers' withdrawal caused severe setbacks for several fighter projects, China still obtained a small number of An-12 turboprop transport aircraft in the 1960s. The PRC Ministry of Aeronautics issued a directive in 1968 to proceed with the development of a licenced version known as Y-8 at the Xi'an Aircraft Company (XAC) and Xi'an's No. 603 Institute, which was responsible for 'large aircraft'.

Development was finalised in 1972 and the first prototype built by using Soviet-made parts made its maiden flight on 15, or possibly 25 December 1974. After this,

This impressive image shows the flight line of the Xi'an Flight College with several Y-7 and Y-8C transports.
(FYJS Forum)

the reverse-engineering was relocated to the newly built aircraft manufacturing complex (known as 012 Base) in Hanzhong, Shaanxi Province. There the first SAC-built Y-8 (prototype 02) flew almost a year later on 20 December 1975, followed by prototype 03 in January 1977. The Y-8 was certified in February 1980 after a brief test-flight phase and serial production started in 1981. In contrast to the original An-12 the Y-8 features a longer, more pointed glass nose as well as a tail station, both derived from the H-6 bomber design. Other changes included a roller-type dropping device instead of a conveyor belt as well as a gaseous oxygen system as opposed to a liquid system. Also, although early production variants of the Y-8 inherited the tail turret with its characteristic twin 23mm cannon, this was removed on later variants. In this form, the Chinese An-12 has now been in production for more than 35 years and, with the latest developments just entering flight status, an end to its service is far from being in sight.

Following a few early production models simply designated Y-8 and dedicated special freighters such as the Y-8A modified to carry the S-70C to Tibet and a civil Y-8B, in 1984 the PLA requested the development of a fully pressurised variant. Since this period falls into the brief Sino-US honeymoon phase, these efforts were assisted by Lockheed (now Lockheed Martin) and led finally to what is currently the most numerous PLAAF model, the Y-8C. The first of these was completed in July 1990 and made its maiden flight on 17 December of that year. Additionally, the original inward-opening two-piece cargo loading doors were replaced by a single flat loading ramp similar to that of the C-130. Lockheed reportedly provided not only assistance with the aircraft's design but also assigned test pilots for the flight tests. In its standard configuration, the Y-8 is capable of carrying 20 tons of cargo, approximately 96 soldiers, or about 82 paratroopers in its cargo compartment which is 13.5m (44.3ft) long, 3m (9.8ft) wide and 2.4m (7.9ft) high. In the medevac role it can carry 60 wounded soldiers with their stretchers or 20 injured soldiers plus three medical attendants.

The Y-8C is the most widely used transport in PLAAF service and the Naval Aviation also operates a few of these versatile tactical transport variants. Based on that version Shaanxi has since developed several special versions like the Y-8H aerial survey and photography variant, the Y-8E drone carrier and several special-mission types which will be described later. Finally, the Y-8 evolved through the improved Y-8F400 (Category II), which was not adopted in PLA service as a freighter but formed the basis for the Y-8GX-3/-4/-7 special mission aircraft up to the Y-8F600 (Category III) which ultimately became the Y-9.

The Y-8C is an interesting evolution of the original An-12, and is still the most important tactical transport in service. This example assigned to the 'fixed wing aviation brigade' of the Airborne Forces.
(CDF)

Shaanxi (SAAC) Y-9

The Y-9 is the PLAAF's next generation medium-size and medium-range tactical transport aircraft to supersede the venerable Y-8 and was first unveiled at the 11th Beijing Airshow in September 2005 as the so-called Y-8X project.

In its original form, the Y-8F600 (the new designation of Y-9 for the domestic version with Chinese systems and powerplants) was a product of a close cooperation between Shaanxi, the Ukrainian Antonov Aeronautical Scientific-Technical Complex (ASTC) and Pratt and Whitney Canada (P&WC) as well as other Western avionics manufactures. The development programme began in 1999 and the two foreign partners joined the programme between 2000 and 2002. Under the original contract signed in 2002, both aircraft manufacturers were jointly responsible for the design, assembly of the prototypes, certification and the initiation of the serial manufacturing. The aim was to completely overhaul, modernise and redesign the Y-8, which resulted in the (probably) final 'Cub' version. The wings and fuselage were redesigned, thus allowing this latest Y-8 version to carry almost a 50 per cent increase in fuel capacity. The Western firms were responsible only for providing the P&WC PW150B engines and the avionics.

For the military, development of the Y-9 began in earnest in October 2005 and construction of the first prototype began in 2006. However, due to the higher demand in the special mission GX series as well as standard Y-8C transport aircraft, the whole project appeared to have been halted in 2007. Both models share the same airframe, but while the transport version received the new Y-9 designation (initially dubbed Y-8U), most EW and special mission aircraft still wear the older Y-8 number. Finally, in late 2008, development was resumed, and the first prototype numbered '741' was rumoured to have made its maiden flight on 5 November 2010.

Successor to the Y-8C is the Y-9, currently in service within only one PLAAF transport regiment. However, production is continuing at a high tempo. (Daniele Faccioli)

The Y-9s assigned to the 4th Transport Division, 10th AR closely support the PLAAF's Airborne Forces in many military exercises. They drop heavy equipment such as armoured fighting vehicles as well as paratroopers.
(Yang Pan via chinamil.com)

In contrast to its ancestor, it features a modified stretched cargo hold capable of carrying a payload of up to 20 tons and its cargo can include a range of military vehicles, helicopters, cargo containers or pallets (13 one-metre sized pallets, 3 four-metre pallets or one six-metre pallet). The enlarged cargo bay measures 16.2m (53.1ft) in length, has a width of 3.2m (10.5ft) and a height of 2.35m (7.7ft) and the aircraft can air-drop paratroops and equipment either by parachute or gravity extraction in either single (up to 8.2 tons) or multiple air-drops (up to a total of 13.2 tons), or 98 armed paratroopers. Alternatively, it can transport 72 seriously wounded patients plus three medical staff or 98 slightly wounded patients. The internal fuel load is 23 tons. It is equipped with a modern four-crew cockpit featuring six colour MFDs, EFIS, advanced communications, a navigation radar behind a new solid nose and, in a strangely shaped tail cone housing, a collision avoidance system to ensure safe flight under all weather conditions. Since both current production models – the Y-8GX series (Gao Xin meaning High New) and the Y-9 – are based on the original Y-8F-600 Category III Platform they share the same uprated WJ-6C turboprops (5,100ehp) fitted with JL-4 six-blade propellers made of composite materials. Both also share small vertical stabilisers installed on the horizontal tailplanes to improve stability at low speeds. Additional electronic equipment include RWR antennas and an EO turret containing FLIR/TV mounted underneath the nose for all-weather/low-altitude operation. The maximum take-off weight is 65 tons, maximum payload is 20 tons or 106 paratroopers, 15-ton payload range is 2,200km (1,367 miles), maximum range is 5,000km (3,107 miles), maximum level speed is 650km/h (404mph), cruising speed 550km/h (342mph), service ceiling 10,100m (33,136ft) and cruise altitude 8,000m (26,247ft).

Following a test phase of about two years, the first low-rate initial production aircraft were delivered to the PLAAF in 2012, while in December 2016 the PLA Army Aviation received its first Y-9 and therefore it appears highly likely that the Naval Aviation will also replace their Y-8Cs in due course. Finally, besides the different Y-9 variants, a new mediumweight turboprop transport aircraft has been under development since May 2017. Known as the Y-30, work is under way at the Shaanxi Aircraft Corporation. This type, first unveiled in model form at the 2014 Zhuhai Airshow, is slightly larger than the Y-9 but smaller than the Y-20 and comparable in configuration to the Airbus A400M.

Harbin (HAIG) Y-12

(ASCC 'Chan')

Development of the Y-12 light transport and utility aircraft began in the late 1970s as the Y-11T. In contrast to its predecessor it featured several improvements including a redesigned wing and a larger fuselage. It first flew in 1980 but only the revised version Y-12II which flew in 1984, updated with more powerful engines and other improvements, became a success. For local military use the standard P&WC PT6A-27 turboprops were replaced by the indigenous WJ-9s from the Y-12IV variant onwards. In PLAAF service this variant is sometimes known as Y-12C and is used for aerial survey missions and parachute training roles. This version is also in operational use by the China Marine Surveillance (CMS). From 2013 another variant, the Y-12D, was introduced for the Airborne Forces, featuring improved WJ-9B engines, as well as military avionics including VHF/UHF radio, INS/GPS navigation, IFF and a FLIR.

Xi'an (XAC) Y-20 'Kunpeng'

The Y-20 is not only by far the largest indigenous Chinese aircraft, but it is also China's first domestically developed heavy transport aircraft. As such, and, in a similar way to the J-20, it represents both significant progress in China's goal of building true strategic airpower and an important milestone for the Chinese aviation industry. Officially,

The Airborne Forces operate several Y-12Ds, especially for parachute training. The type is slowly replacing the old Y-5 as a small and versatile transport. (FYJS Forum)

the Y-20 is named the Kunpeng after a giant mythological bird; in PLAAF service it is nicknamed 'chubby girl'.

The initiation of the Y-20's development was closely connected with China's failed acquisition of additional Ilyushin Il-76MD transports and Il-78 tankers making the design of an indigenous military transport a high-priority project. Naturally, the development of such a complex aircraft was a demanding exercise for the Chinese aviation industry and although the full background has not entirely been made public, it seems that in early April 2006, Chinese representatives approached the Ukrainian company ANTK Antonov – by then involved as a consultant in the development of the Y-9 transport and ARJ-21 airliner – to assist in its development. The true extent of Antonov's involvement in what became the Y-20 was much more than consultation. It contributed considerably to the aircraft's development, reportedly with a jet-powered derivative of its An-70. On the Chinese side, the Xi'an Aircraft Company (XAC) was given responsibility for the project and the design team was led by Chief Designer Tang Changhong at XAC's No. 603 Institute. According to reports, in the following three years the design was revised, due to changed requirements from the PLAAF for an increased maximum take-off weight and dramatically higher payload. The whole project switched to a substantially larger and heavier proposal with a take-off weight of 230,000kg (507,063lb) and a maximum payload of 60,000kg (132,277lb) in order to be able to carry the PLA's latest MBT, the Type 99A2, which has a weight of approximately 58 tons. In this form, the design received its Y-20 designation in late 2009. At around the same time the first concept drawings appeared on the Chinese internet, stating that the prototypes and

2018 saw the first widespread use of the Y-20A in cooperation with the Airborne Forces. This aircraft, 11057, carries the highest serial number so far confirmed.
(via chinamil.com)

early serial aircraft were to be powered by the Russian D-30KP-2 engines, with later production versions using a new indigenous high-bypass turbofan called the WS-20 Huanghe. Between 2008 and 2010 several full-scale mock-ups of the aircraft's sections were built and by mid-2009 construction of the first prototype had begun. Taxi tests began on 21 December 2012 and a first grainy image of the prototype appeared on the Chinese internet on Christmas Eve 2012 at the CFTE airfield at Xi'an-Yanliang. The maiden flight was successfully completed on 26 January 2013.

Technically the Y-20 follows the typical military transport layout, but it is clearly a unique addition to the family of other heavy transport aircraft in fuselage shape, landing gear empennage and wing design. Comparable in size to the Il-76, the Y-20 is actually 2-3m shorter than Il-76 but with a wingspan roughly the same at around 50m (164ft). However, the two most significant differences are the wider fuselage and the wing. Compared to the 4.8m of the Il-76, the Y-20 uses a wider and taller fuselage of 5.5m, which offers a more flexible cargo hold of greater volume, making it more versatile for carrying large-sized cargoes, one of the main requirements of the PLA. Also, the Y-20 reportedly incorporates a higher proportion of modern materials and is fitted with a more modern digital cockpit, flight instrumentation for a crew of three and a digital FBW system. Besides that, the Y-20 is equipped with a small FLIR (enhanced vision system, EVS) below the forward windshield to assist taking off and landing under poor weather conditions and has a dorsal SATCOM antenna behind the wings. Another important item is a modern supercritical aerofoil section of the wing which features a slightly higher aspect ratio and is less swept on the trailing section in comparison to the Il-76. This presumably gives the aircraft better fuel economy thus further extending its range. However, the Y-20's high-lift devices along the wing leading and trailing edges are surprisingly elaborate, featuring a heavier triple-slotted flap, more akin to the Il-76 than to the C-17. Reports quote the Y-20 as having a maximum payload of 66 metric tons and a maximum take-off weight of more than 200 tons; other sources state an expected payload of 50 to 55 tons and a maximum take off weight of 180 to 200 tons, depending on the type of engine is powering the aircraft. The engines are the current weak point in the design, especially in comparison with Russia's Il-476, since the Y-20A still has to rely on dated D-30KP-2s, which lack the thrust and efficiency of modern turbofans. According to the Russian media, a first contract between Rosoboronexport and China was signed in April 2009 for 55 engines, which were all to be delivered by 2012. An additional second order was signed in late 2011 with NPO Saturn for an additional 184 D-30KM-2 engines, to be delivered within the next four years and to be used on the Y-20 transports as well as the H-6K bomber version.

Following the prototype number 781, altogether five more prototypes followed until service entry in June 2016. In December 2013 the second aircraft, number 783, flew for the first time, followed by two or three more (numbers 785, 786? and 788) during 2015. The final prototype, number 789, already in low-rate initial production (LRIP) standard, flew in February 2016. In late 2015 it was officially announced that the research and development (R&D) phase of the Y-20 programme was completed and serial production on a newly established pulsed assembly line had begun. Official handover to the PLAAF took place on 15 June 2016, and on 6 July it was officially introduced to PLAAF service as the Y-20A. By May 2018, a total of seven had been assigned to the 4th Transport Division and several more are ready for delivery. Altogether, the PLAAF's requirement is, according to a military expert interviewed by the People's

Daily Online, for no more than 100 in the future with a few more eventually as special mission aircraft such as AEW&C aircraft or tankers.

The future

For the time being, the biggest weakness in its current form is once again the lack of a modern high-bypass-ratio turbofan engine, which severely limits its performance until the 'final version' appears. As stated earlier, this new engine is the WS-20 Huanghe, which will be China's first modern high-bypass turbofan. It is based on the core of the military WS-10A, as used in the J-11B and J-16 fighters. With a projected thrust range of 120 to 140kN, the WS-20 entered testing in 2014. Testing has continued at the CFTE using a converted Il-76LL engine testbed in parallel with the Y-20's service introduction and LRIP. Therefore, it seems as if – comparable with the J-20A – the PLAAF is more interested in introducing a first batch of this type, even if not up to full performance, as soon as possible, in order to expand its transport capabilities and then to proceed with the Y-20B with the WS-20 engines. A first glimpse of how this definitive version will look was made apparent at the Airshow China in 2014, where a civilianised version called the Y-20-F100 was presented. According to the latest reports, progress on the WS-20 engine is proceeding well and the maiden flight of a Y-20 with these engines is expected for 2019. For the future, several special-mission aircraft are planned based around the Y-20 such as the Y-20U tanker and the KJ-3000 AEW.

The current Y-20A is limited in its capabilities due to its reliance on old D-30KP-2 engines, but the first Y-20B powered by the indigenous WS-20 is reportedly close to being unveiled. This concept of a civilian Y-20F-100 shows how it might look like. (CDF)

Foreign types

Airbus A319 ACJ

Besides several indigenous types, the PLAAF's VIP fleet comprises a number of Western aircraft including three Airbus A319-115 ACJ airliners. Originally delivered to China United Airlines in full CUA livery, they were later transferred (in September 2013, August 2014 and January 2015 and numbered B-4090, 4091 and 4092) to the 34th Division as VIP transport aircraft. For several years, they have been formally painted with full PLAAF insignias, similar to those of the Boeing 737-800 VIP transport operated by the same unit.

Boeing 737-800 BBJ

Before the A319 was introduced, the PLAAF flew several Boeing 737s. Altogether three different variants are in service: eight 737-3Q8s (B-4008, 4009, 4018 to 4021 and B-4052 4053), two 737-76Ds (B-4025 and 4026) and four 737-85Ns (B-4080 to 4083). These too, are operated by the 34th Division as VIP transport aircraft and were again initially painted with the CAU livery but are now formally painted with PLAAF insignias. Two 737-300s were converted to command posts.

Bombardier CRJ200 and CRJ700

Complementing these larger single-aisle airliners are five Bombardier Canadair CRJ100 Series (CRJ200) – CL-600-2B19 Challenger 600s – regional airliners (B-4005, 4006, 4007, 4010 and 4011), which were introduced between September 1997 and June 1988 and 10 Canadair CRJ-700 Series – CL-600-2C10 Challenger 870 – (B-4060 to 4069), which were acquired between January 2005 and February 2015.

Besides the Boeing 737, the PLAAF also operates three A319-115 airliners as dedicated VIP transports. (SAM-L00Lin-t-n-t via Top.81 Forum)

For many years a handful of Boeing 737 variants have operated as VIP transports within the 100th AR. This particular aircraft is a Boeing 737-76D (c/n 33472) and was delivered in September 2003 after being transferred from Shanghai Airlines as B-2689. (Sunshydl)

Ilyushin Il-76

(ASCC 'Candid')

For many years the Il-76 was the PLAAF's sole multi-purpose strategic airlifter. Designed by the then Soviet Ilyushin design bureau it first flew in March 1971 and entered service within the Soviet Air Force in June 1974 to become the main Soviet strategic transport aircraft. The first Il-76 for the PLAAF were purchased via CAAC, when, between in 1991 and 1996, 14 Il-76MDs were delivered to China in two batches by the Chkalov Tashkent Aircraft Production Company (TAPOiCH), an Uzbek aircraft manufacturer. Interestingly, these first aircraft appear to be unarmed Il-76TD models with the tail-gun and other military equipment removed and initially they were operated by the China United Airline (B-403x -404x) in civil markings. Only later were they officially transferred to the 13th Transport Division to regularly support the PLAAF's Airborne Forces.

For the PLAAF, the Il-76 became the first true strategic transport to significantly boost the PLA's rapid reacting and long-range airlifting possibilities by virtue of its impressive capabilities (maximum load 48 tons, normal range 5,000km/3,107 miles) where they are used to transport not only paratroopers, but also heavy equipment including up to three ZBD-03 AFVs. Due to the small numbers acquired – including four which were converted to the KJ-2000 AEW – several more Il-76MDs as well as Il-78 tankers were expected to be purchased. Although a contract for 34 Il-76MDs and four Il-78Ms, worth a total of USD1.045 billion was signed between Rosoboronexport and China's National Defence Ministry in September 2005, this most important deal never materialised. There were several reasons, both political and industrial, but eventually, after the break-up of the former Soviet Union the Tashkent-based company was simply unable to resume production.

After several attempts to renegotiate the terms of the contract both in price and number of aircraft the contract was finally cancelled in March 2006. As a consequence, the design of a '200-ton military aircraft' – as the Y-20 was unofficially called at that time – gained the status of a high-priority project in the same year. In the meantime, China acquired several more ex-USSR Il-76MD/TDs on the second-hand market and overhauled them in Russia as a stopgap measure until the Y-20 entered service. A first batch comprised three Il-76s from Russia in December 2011 and all were delivered in 2012. These included at least one Il-76TD (number 21141) and a second contract was announced by the Russian military export agency Rosoboronexport in June 2013 for 12 Il-76MD aircraft. In summary, the PLAAF operates 12 Il-76MDs, two more unconfirmed TDs or MDs and seven Il-76DTs with one more TD on delivery in the transport role and all are expected to be complemented with and later replaced by the indigenous Y-20A.

As with the Russian Air Force, the Il-76MD is the workhorse for moving heavy equipment and long-range duties, but available in limited numbers. It is now in service with two regiments within the 13th Transport Division due to acquisition of additional aircraft via the open market. (Teerawut Wongdee)

Ilyushin Il-78

(ASCC 'Midas')

Originally, it was planned to purchase four Il-78 tankers, which were included in the contract signed in 2005. However, after that deal fell through, the PLAAF had again to negotiate on the second-hand market for a substitute. This was especially pressing since the PLAAF's Su-30MKKs were not compatible with the HU-6 tankers. By this time, although unconfirmed, several reports claimed that a contract was signed with Ukraine in 2011 for three ex-USSR Il-78s, which were to be refurbished by the Nikolaev

Complementing the Il-76 transports, three Il-78 tankers were acquired to support the Su-30MKK fighters, which were not compatible with the HU-6 tankers. (CMA)

Aircraft Repair Plant. A first one – claimed to be an ex-Ukrainian Air Force Il-78 – was noted in the characteristic PLAAF colour scheme in March 2014. The first Il-78 was confirmed in China in October 2014 and since then at least two more have been delivered to the PLAAF 13th Division (numbers 20641 and 20643) in June 2015 and April 2016 respectively. Allegedly, at least one more airframe is due to be delivered.

In a similar way to its Russian armed forces equivalents, in PLAAF service they carry three UPAZ-1A refuelling pods – although some of the rare images suggest that the third pod on the port side of the rear fuselage is not always installed – and are able to carry around 60 tons of fuel, giving a range of 7,300km (4,536 miles). They have been seen supporting Su-30MKKs during their long-range patrol missions over the South China Sea and the East Pacific near Japan. In the future they are expected also to support J-16s and Su-35s until an indigenous development based on the Y-20 enters service.

Tupolev Tu-154M

(ASCC 'Careless')

Although it is often reported that the PLAAF's 'regular' Tu-154M civil airliner has been retired or converted to Tu-154M/D special mission standard, four examples are still in use. The type entered PLAAF service in the mid-1980s and a maximum of nine were in use. Several have been converted in recent years or replaced by 737s and A319s.

Support, transport and surveillance aircraft currently in use with the PLAAF

Type	Role	Service entry	No (est.)
An-24/An-26	Tactical transport	1977?	n/k
An-30	Aerial survey	1978?	4-6
A319-115	VIP transport	2013	3
737-300	VIP transport	1988	6
737-700	VIP transport	2004	2
737-800	VIP transport	2010	4
CRJ200	VIP transport	2001/02	5
CRJ700	VIP transport	2014	14
HU-6	Tanker	mid-1990s	12
Il-76MD/TD	Strategic transport	1991	24
Il-78	Tanker	2014	3 or 4
Tu-154M	VIP transport	mid-1980s	4
Y-5C	Transport, liaison	1960s	some 200?
Y-7	Tactical transport	1984	some 60
Y-7H	Tactical transport	1992	some 80
Y-7G	VIP transport	2014	6
Y-8C	Tactical transport	mid-1990s	some 84
Y-8H	Aerial survey	mid-1990s	2-4
Y-9	Tactical transport	2012	some 30, i/p
Y-12D	Aerial survey	2015	30
Y-20A	Strategic transport	2016	some 10, i/p

Special mission aircraft

The PLAAF operates a wide range of specialised versions of the Y-8, several of which are now somewhat dated while others are just being introduced. Understandably, little is known about the exact nature of their missions or their individual equipment. Additionally, and in contrast to the previous versions – the different aircraft are no longer designated in a strict sequential order but sometimes simply by a letter denoting the role of that type. The future of this versatile transport and its special mission offsprings would seem to be secure for many years to come.

Special Mission Aircraft of the Gao Xin (High New) series

Following the Y-8X and Y-8J the Y-8 evolved within the PLAAF and Naval Aviation into a similar mainstay for special roles similar to those of the C-130 or C-135 for the US and since the early 2000s several more special-purpose variants were fielded. Their missions ranged from electronic warfare (EW), electronic intelligence (ELINT), signals intelligence (SIGINT), command posts, offensive electronic countermeasures (ECM), early warning and surveillance as well as anti-submarine and psychological operations. These are all known to be initiated under the so called Gao Xin – meaning High New – project. Understandably, only scant information is available about the history of individual versions or even their specific mission equipment and, in most cases, little was known about these projects until a new Y-8GX version was first seen. Although these Y-8GX versions are sequentially numbered from GX-1 to GX-12, they were not seen in the same order. However, in order to give a clearer summary of the different versions, they will be covered here in that numerical order.

The first of these models was seen as operational in 2004 and there are persistant rumours suggesting that some of the equipment was derived from a US Navy EP-3 ELINT aircraft, which made an emergency landing in Hainan Island in April 2001 after colliding with a Chinese J-8 fighter. Even assuming the Chinese inspected this aircraft and gained an insight into its equipment, it seems unlikely that within three years the EP-3's electronics were 'copied'. Indeed, bearing in mind China's already impressive achievements with the KJ-2000 system and the country's booming electronic and telecommunications industries, it is more likely that the PLA was already working on such systems and was well capable of developing its own indigenous EW/ELINT systems. Although perhaps less effective than the latest Western systems these types are doubtless equipped with an extensive array of sophisticated intelligence-gathering equipment to either monitor enemy electronic activities or to be capable of launching offensive jamming against enemy communications and radar systems.

Y-8CA Radar Testbed
After the first two dedicated specialised mission types, a single Y-8C ('079') was converted by XAC into an airborne radar testbed to test various radars and other avionics. The aircraft is known simply as 'Radar Electronic Testbed Aircraft' and is operated by the China Flight Test Establishment (CFTE) based at Xian-Yanliang, Shaanxi Province.

Its most noticeable feature is a modular radome section which replaced the original Tu-16-style front section and is capable of mating with radomes of different sizes

and shapes (this aircraft has been seen with both pointed and blunt noses) depending the type of radar being tested. Following its conversion, not only was the new front section added but also the rear ramp was removed to make full use of the full pressurised cabin serving as a flying laboratory for the test engineers. This particular aircraft was introduced at the CFTE in August 1999 and was reported to be involved in the development and testing of fire-control radar systems for most of China's latest generation combat aircraft including the J-10 (the radar was also seen on a Y-7 testbed), the J-11B and JH-7. Following the latest sightings, it is currently equipped with the PESA (or AESA) radar for the J-10B and yet another testbed is reportedly testing the J-20's avionics. These latest modifications also include a huge canard-like fairing on the fuselage.

The Y-8CA has not only acted as a radar testbed, but it also served as the testbed for the new JL-4 six-blade high-efficiency propeller installed on its WJ-6C1 turboprop engine. The new propeller has been adopted by the Y-8 Type III platform and Y-9.

Y-8CB = Y-8GX-1 (High New 1) ELINT

(ASCC 'Maid')

Following the retirement of the HD-5 EW/ECM aircraft in the early 1990s, the PLAAF required a more modern replacement for the jamming role, which was seen for the first time in Nanjing, Jiangshu Province in July 2005. Following unconfirmed reports, that version made its maiden flight on 26 January 2000 and currently at least five aircraft (numbers '3001x' plus '30511') are operational with PLAAF.

The Y-8CB (GX1) is one of the original early-generation special mission aircraft still based on the old Y-8C or Category I airframe. It is a dedicated ELINT variant and will likely be replaced by the new Y-8GX-12. (Top.81 Forum)

In contrast to the Y-8GB the Y-8G (GX-3) uses the slightly more modern Category II platform fearing the revised front section of the Y-8F-400. This type is expected to be replaced in its ECM role by the latest Y-8GX-11.
(DS via Top.81 Forum)

The most striking features of the Y-8G are the characteristic 'hamster' cheeks, which most likely contain a very large ECM antenna array for long-range electronic/communication jamming purposes.
(Longshi via CDF)

This version, called Y-8CB and also known as Y-8GX-1 (K/JYZ-8), is based on the Y-8 Category I Platform (also known as Y-8C) which features the same 'sealed' rear loading ramp, which is crowded by an array of antennas protruding from it. Several more antennas are located under the fuselage between the main landing gear fairings and around a ventral canoe fairing underneath the forward fuselage which may house a large ECM or SAR antenna.

Latest images indicate that some aircraft have been upgraded with an additional dorsal SATCOM antenna in front of the vertical tail. Its main equipment is reportedly the BM/KZ300 and 308 used to jam enemy communication networks and radar systems. In recent years they have been further upgraded with an additional datalink equipment and a dorsal SATCOM antenna.

Y-8G = Y-8GX-3 (High New 3) ECM/ELINT

Although the previous Y-8GX versions had been heavily modified, they appeared externally to be more or less standard Y-8s. This, however, changed with the Y-8G – also known as Y-8GX-3 (K/JYG8) – which is based on the Y-8F400 Category II Platform with rear cargo loading ramp and tail turret removed. The Y-8G was accidently unveiled during the Chinese Vice Prime Minister's visit to SAC in April 2005 and reportedly had its maiden flight in late 2004. Altogether five Y-8GX-3s ('30x1x') have been in service with PLAAF for a few years with three more having entered service recently ('2077x') although one was lost in January 2018.

This version features two huge 'hamster' cheek fairings behind the cockpit section and in front of the wings, similar to, but much shorter than those of the US RC-135V/W and which reportedly house an ECM antenna array for long-range electronic jamming. This system – allegedly a product of the No. 14 Institute – could also provide data for long-range battlefield surveillance. Other noticeable features are a redesigned solid nose with the typical undernose radome removed and a large tail fairing to provide 360-degree coverage and a cylindrical fairing on top of the vertical tail fin related to its ELINT system. The Y-8G is expected to be replaced by the new Y-9G (GX-11).

Y-8T = Y-8GX-4 (High New 4) C3I Airborne Command Post

The next in line is the Y-8T – also known as the Y-8GX-4 – which was seen for the first time at the China Flight Test Establishment (CFTE) based at Xian-Yanliang, Shaanxi Province, reportedly after completing its maiden flight in August 2004.

This version is operated as a dedicated C3I airborne command and control post to provide better coordination for PLAAF air operations and at least five – and possibly six – aircraft are in service (numbers 30871 to 30876) with PLAAF.

As is the Y-8G, the Y-8T is also based on the Y-8F400 Category II Platform. However, this aircraft features a rear-facing radome the purpose of which is unknown (but possibly for satellite communications) and a dorsal fairing aft of the wing section which possibly houses a SATCOM antenna. Additionally, there are several antenna arrays along

The Y-8T (GX4) is a dedicated C3I airborne command and control post to coordinate PLAAF operations at theatre command level.
(FYJS Forum)

the top and bottom of the fuselage, as well as on the vertical tail fin which together indicate a comprehensive communications and sensor suite.

Shaanxi Y-8W/KJ-200 = Y-8GX-5 (High New) 'Balance Beam' AEW

(ASCC 'Moth')

The Y-8W or KJ-200 (K/JE03) is China's second and smaller tactical AEW type following the large KJ-2000. If it was to be one of two indigenous fall-back options should the A-50I purchase fail or, if it was intended as a smaller tactical complement to the huge but expensive KJ-2000, is still not clear, although hints that such a type was under consideration have appeared on the net since the mid-1990s. Regardless of its precise origins, the KJ-200 has been developed by the Shaanxi Aircraft Industry (Group) Co. It is believed that its development began during the late 1990s, from the outset being intentionally based on the new Y-8F-600 or Category III transport. At first an older Y-8F-200 prototype (B-576L) was converted into a radar testbed and was seen in October 2004 at the CFTE undergoing extensive modifications featuring most significantly a strut-mounted antenna of a phased-array ESA-type AEW radar similar to that of the Swedish Ericsson Erieye. It was later revealed that this first prototype had made its first flight on 8 November 2001. Because of its very distinctive appearance this aircraft gained the unofficial nickname of 'Balance Beam'. Additional radomes are located at the nose tip, under the nose, on the wingtips, in the tail cone and on top of the tail fin, most likely to provide full 360-degree coverage. The radar itself was probably developed by the 38th Research Institute at Hefei and was initially tested on a Y-7.

In contrast to all previous EW variants, the KJ-200 was the first to use the improved Cat III platform from prototype '02' which has a redesigned fuselage with a solid nose

The Y-8XZ (GX-7) is again based on the Category II platform and is a dedicated psychological warfare variant recognisable by large fairings located forward of the main landing gear compartments. (CMA)

and a new tail section with the loading ramp removed. It has a new glass cockpit and two radomes are located at the nose tip and tail cone which might house additional AEW antennas. More fairings can be seen at the wingtips and on top of the tail fin housing ESM antennas and a series of small antennas are located on top of the forward fuselage. It also has an integrated wing fuel tank and four high-efficiency JL-4 six-blade propellers giving the aircraft a longer range (5,000km/3,107 miles) and producing less noise. Its C3I centre is composed of around eight display consoles in a large pressurised cabin. A new integrated digital avionics system based on the ARINC429 and RS422 data bus has been installed.

The second prototype first flew on 14 January 2005. However, it had only a brief career, since it was lost during a crash on 4 June 2006, causing the development programme to be temporarily halted. After a delay of a year, which included some redesign, the production resumed. The most important measures were a strengthened fuselage and the attachment of additional small vertical stabilisers to the tips of its horizontal tail fins. The project itself gained momentum during the following years and the PLAAF introduced a total of five Y-8Ws and, in addition, the Naval Aviation has taken delivery of six Y-8WHs, also known as KJ-200Hs or HJ-200s ('9371', '9381', '9391', '9401', '9411' and '9421'). In December 2016 an image appeared showing that at least one KJ-200 with the PLAAF and Naval Aviation had been modified with a new nose to replace the typical 'Pinocchio' nose. It is said to cover an AEW radar antenna, which could give the aircraft a better coverage in the forward hemisphere and, as a result, the chin-mounted weather radar has been removed. Currently, it seems likely that all remaining KJ-200s will be upgraded to the KJ-200A standard.

Y-8XZ = Y-8GX-7 ('High New 7') Psychological Warfare Variant
Yet another unusual version is the next-in-line Y-8GX-7, which is also known the Y-8XZ (K/YXZ8) and is acting as a psychological operation variant similar to the US Air Force EC-130E Rivet Rider or EC-130E Commando Solo. Little is known of this type but at least two aircraft (3101x) have been in PLAAF service since 2007, since it was unveiled in April 2008. In contrast to the last two GX versions above, it is again based on the older Y-8F400 with loading ramp and tail turret removed and can be identified by two huge fairings located forward of the main landing gear compartments with prominent dielectric openings facing to the sides. There are two large plate antennas on each side of the rear fuselage and two blade antennas on both sides of the vertical tail fin. Additionally, they feature a rear-facing radar similar to that on the GX-3 and GX-4 as well as a wire antenna underneath the rear fuselage and a prominent SATCOM antenna on top of the rear fuselage. Regarding its mission fit, it is reportedly equipped with high-power broadcast equipment covering AM, FM, SW, TV plus various civilian and military communication bands, abling it to jam enemy communications as well as disrupt and demoralise an enemy with propaganda broadcastings. According to the latest information, the Y-8XZ is expected to be replaced by the new Y-9XZ (GX-9).

Successor for the Y-8XZ is the more modern Y-9XZ (GX-9), which shares a similar antenna fit but is based on the latest Y-9 airframe. (meyet.com)

Shaanxi Y-9XZ = Y-8GX-9 (High New 9) ELINT
Little is known of this variant, which was at first thought to be a re-engine programme for the Y-8XZ (GX-7), but in October 2012 it became clear that this new psychological warfare variant (High New 9/Y-9XZ?) is in fact a new development based on the Y-9 platform. The aircraft was rumored to have a new capability of hacking into enemy

The KJ-500 was an ambitious project for a more capable medium-sized AEW type, that could complement the larger but numerically limited KJ-2000. At least eight are operational within the 26th Specialised Division.
(Yang Jun via chinamil.com)

communication networks. Therefore, it could be used to interrupt the internet traffic or spread false information and create chaos through social networks within the enemy society by hacking into key web servers. It has rarely been seen in operation but it is believed that several Y-9XZs have entered service with the PLAAF since mid-2014 ('30x1x?') replacing the earlier Y-8XZ (GX-7) in a role similar to that of the USAF's EC-130J Commando Solo.

Shaanxi Y-9W/KJ-500 = Y-8GX-10 (High New 10) AEW

This type has been under development at SAAC since the late 2000s and will form the next generation of medium-sized AEW&C platform. In contrast to its predecessor, the KJ-200, the KJ-500 had its characteristic 'balance beam' radar replaced by a more traditional fixed rotodome containing three AESA arrays arranged in a triangular configuration similar to that of the KJ-2000. The radar is allegedly a new system and a product of the No. 38 Institute which utilises the latest digital radar technology. During its development several radars and different shapes of the radome were evaluated on the Y-8CE testbed and besides the three main antennas an additional SATCOM antenna was integrated to the top of the rotodome.

Similar to other members of the EW family, this variant also has an enlarged nose and tail radomes which could house additional radar antennas to cover both forward and rear hemispheres and, akin to the Y-9JB, it features the same two rectangular bar-shaped fairings housing ELINT antennas on both sides of the rear fuselage. Additionally, it is equipped with MAWS sensors installed aft of the cabin door and ahead of the tail. Only two KJ-500 prototypes had been built by late 2013 and the first serial aircraft

A surprising development was the unveiling of the first improved KJ-500A featuring a fixed IFR probe. By May 2018 at least two prototypes were known.
(FYJS Forum)

entered service with the PLAAF at the end of 2014; eight are now operational. In April 2015 images confirmed that the Naval Aviation has also taken delivery of four to six KJ-500H AEW&C aircraft (also known as the HJ-500) and this type is still in production.

Latest variants:
Shaanxi Y-9G = Y-8GX-11 (High New 11) ECM and Y-9X = Y-8GX-12 ELINT

Y-9G (GX-11), which is a new ECM variant similar to original Y-8G (GX-3) with its characteristic 'hamster cheeks' to suppress enemy radar and communications. Following the latest information it is based on the Y-9 platform, features a new chin radome plus three large oval and rectangular-shaped antenna panels along each side of the fuselage. Additionally, it has two plate antennas attached to the vertical tail fin complemented by an array of blade antennas under the fuselage and a small semi-spherical antenna underneath the wingtips. Reportedly two examples have been completed and became operational in early 2018.

The second of this new type is similar in external appearance to the Y-9JZ (GX-8) which is flown by the Naval Aviation. So far it is named Y-9X (to become GX-12) as a placeholder and is rumored to be a new ELINT type for the PLAAF. Only one prototype (CFTE serial number '745') has been seen at the CFTE in Xi'an-Yanliang. If the rumours are correct, it would make sense for the aircraft to be based on the Y-9JZ, since it features a similar nose and two large rectangular-shaped antenna arrays on both sides of its rear fuselage. More antennas are installed all over the fuselage, in the tail cone and on top of the vertical tail fin. Something that could be a SATCOM antenna is also fitted on top of the mid-fuselage. This aircraft could be a replacement of Y-8CB (GX-1) and could become the Y-9CB.

The Y-9G (GX-11) has been known since 2011 but was only seen in operational service during an exercise in July 2018. It is slowly supplementing or even replacing the Y-8G (GX-3). (via East Pedulum)

Foreign Types

Boeing 737-300 Command Post

Complementing the handful of Boeing 737s as VIP transports are two 737-3Q8 airliners (B-4052 and 4053) converted into airborne command posts. The two aircraft were originally purchased from China United Airline in 1990 and were converted without US approval by the Xian Aircraft Corporation (XAC). The most prominent feature is one large fairing on top of the forward fuselage and two small fairings located underneath the mid-section of the fuselage, presumably to house communication and datalink antennas. Following their modification, they initially wore a light grey colour scheme but this was changed to a modified standard PLAAF VIP transport scheme. One aircraft (B-4052) was allegedly used as a missile tracking platform for strategic missile tests.

Not directly assigned to the PLAAF, a few Cessna Citations feature different varying mission equipment configurations including this unknown semi-conformal pod with special antennas. (FYJS Forum)

Learjet Model 35 and Model 36

Also in use for special missions operations are five Learjet Model 35 and 36 business jets, which were acquired in 1985 and 1986. In PLAAF service three Learjet 35As (B-4186 to 4188) and two Learjet 36As (B-4184 to 4185) were converted into specialised ELINT reconnaissance aircraft by the PLAAF as Aerial Remote Sensor Systems (ARSS) with the 102nd Independent AR before the dedicated Y-8 High New series entered service.

Complementing the larger Y-8 special mission types, the PLAAF operates a handful of slightly differing Model 35 and 36 Learjets as specialised ELINT or reconnaissance aircraft.
(Top.81 Forum)

Cessna 550 Citation II and Cessna 650 Citation III

Originally, eight Cessna 550 Citation IIs and two Cessna 650 Citation IIIs were acquired – of which only two are still in service (B-4101 and 4102) – and are flown by the Airborne Remote Sensing Centre (ARSC) assigned to the Chinese Academy of Sciences.

Tupolev Tu-154M/D

(ASCC 'Careless')

Besides the above Western types there are also two Russian types in service. The first is the Tupolev Tu-154M, of which several have been converted into ELINT aircraft. In this configuration they are known as Tu-154M/Ds (Type I/II), and they entered service in 1995. They were originally acquired under the cover of the civilian China United Airline registration and operated to search, locate and analyse radio signals. The original Type I was equipped with a BM/KZ800 ELINT system and all have been updated to

After retirement as VIP transports, the PLAAF converted several Tu-154M into Tu-154M/D special mission aircraft.
(Top.81 Forum)

Type II configuration. Type II accomplished its flight test in 1996 and features a huge canoe-shaped fairing under the fuselage which is believed to house a modern synthetic aperture radar (SAR) to provide high-resolution ground mapping images. In recent years they have also been used to monitor and escort H-6Ks during their missions over the East China Sea. Of a total of 16 acquired by the PLAAF, four were initially converted to Type I, but with the retirement of VIP transports, an additional five airframes were modified together with the first four to Type II, so that now, nine are known (B-4015, 4016, 4017, 4024, 4027, 4028, 4029, 4050 and 4138).

Tupolev Tu-204-120C(E)

One very special type in PLAAF service is a single Tupolev Tu-204-120C(E). This aircraft was first noted at the CFTE in Yanglian in September 2011, still showing the former ACC logo on the tail and it was rumoured to be acting as a tanker testbed. However, it became a dedicated avionics testbed related to the J-20. The aircraft was originally RA-64030 (c/n 1450743664030) built at ZAO Ulyanovsk Aviastar-SP and was powered by Rolls-Royce RB211-535E4B-75 engines. It is the first – and, so far, the only one built to these specifications – from an original contract signed by the bankrupt Egyptian company Sirocco Aerospace International in September 2001. It was later briefly operated by Air China Cargo (October 2008) as B-2871 before it was stored at Tianjin and finally delivered to the CFTE at Yanliang in May 2011. After an analysis of its structure the JSC Tupolev received a contract in late 2012 to convert this aircraft into a flying laboratory to be used by the CFTE. In this form it made its first flight on 11 December 2013, and since then it has carried the CFTE serial '769'. Its most important modifications include the J-20's radar and CNI system and between May 2014 and April 2015 it received further changes including a wing-like structure on top of the front fuselage similar to that of the the Boeing 757 used as a F-22 testbed and it is still at Yanliang.

KJ-2000

(ASCC 'Mainring')

The KJ-2000 is the first and, so far, the largest dedicated AEW aircraft in PLAAF service and it stems from a requirement following the cancellation of the indigenous Y-10 AEW project. In fact, it was originally planned to purchase a Western system as a follow-on to the Y-8J contract. A first proposal was built around the British GEC-Marconi Argus 2000 mechanically scanned AEW system originally planned to be fitted in the BAe Nimrod AEW.Mk 1 but installed onto the Il-76MD in a similar arrangement to that of the Nimrod AEW in two bulbous nose and tail cone radar-arrays. The second proposal, which later won the original request was from the Israeli company Elta Electronics, who offered a more sophisticated system based around the EL/M-2075 Phalcon, which also was originally planned to be fitted in a similar way but complemented by two side-mounted arrays. The political situation severely worsened following the Tiananmen unrests 1989 and, despite the Western embargo, the Israeli project proceeded as planned and in 1992 negotiations with Israel and Russia for a possible purchase of four 'green' Il-76MD aircraft without the radar began. This version of the A-50 was therefore re-designated A-50I or A-50AI. An agreement was signed in May 1997 for one aircraft for USD250 million, plus an option of three more amounting to USD1 billion.

In parallel to the Boeing 757 used as an F-22 testbed, the CFTE acquired this single Tu-204C, which was modified with the J-20's radar and CNI system for additional tests as an avionics testbed. (Top.81 Forum)

The first KJ-2000 prototype was based on a Russian A-50I airframe but equipped with indigenous AEW and C4ISR systems. This is the second of four serial conversions now operational within the 26th Specialised Division. (Jian Feng)

The design has since been revised to provide better radar coverage, with the radar installed in a traditional radome on top of the fuselage. After some delays, on 25 October 1999, Russia delivered the first A-50I prototype with the civil registration RA-78740 (c/n 0093486579) – actually a former Russian military A-50 – to Tel Aviv for the installation of the Phalcon system.

However, the installation was not completed, due to an increasing political dispute between the US administration and the Israeli government, so that in July 2000 the contract was cancelled and the installed equipment was removed. Consequently, China had to continue with its own efforts to develop a modern operational AEW system – which were continued in parallel – alias Project 998 at the No. 603 Institute, the Xi'an Aircraft Company (XAC) and the Nanjing Research Institute of Electronic Technology/No. 14 Institute, who were responsible for the radar. After the Israeli systems were retrieved, the airframe was handed over to China in 2002, where Xi'an began with the airframe modifications and installation of the indigenous Type 88 radar and a C4ISR system. This is not a single radar but three AESA arrays in a triangular configuration mounted within a fixed rotodome. Its maximum range is said to be 470km (292 miles). In contrast to the Russian A-50, the Chinese rotodome is both larger in diameter and width and is distinguishable from the original A-50's design by revised mountings. Other differences are a solid nose replacing the original 'glass' section and

the huge horizontal strakes along the main-gear fairings have been removed. Instead of these the KJ-2000 features two canted trapesoidal fins near the fuselage end, and two semicircular dielectric devices on the outer wingtips of unknown design. Another feature is a SATCOM/SATNAV blister ahead of the wing on top of the fuselage which is of a slightly more bulbous design than the Russian equivalent. The KJ-2000 has a regular flight crew of five and, following some information, 11 mission crew members. It can survey for about 12 hours and has a maximum range of 5,500km (3,418 miles).

The first KJ-2000 prototype was renumbered as '762' – still in Russian military configuration plus an IFR probe – and made its maiden flight on 11 November 2003. Flight testing continued until late 2005. In the meantime, four additional Il-76MD transports (B-4040 to 4043) were diverted for conversion and delivered to the PLAAF. They were declared operational in December 2007. The original A-50I prototype ('762') had its rotodome removed in late 2014 and was converted into an engine testbed for the WS-18 turbofan – an indigenous development of the Russian D-30KP-2 engine and, in addition, the PLAAF operates yet another A-50-derived airframe as a dedicated engine testbed. This particular aircraft is a former Russian Air Force Il-76SKIP/Be-976 missile tracking aircraft (RA-76456), which was acquired by China in mid-2005 following the removal of the radar system at the LII Flight Research Institute. To act as an engine testbed, the inner port side engine has been removed and replaced by an arrangement similar to that of the Russian Il-76LL variant. Two large wingtip pods were retained, and two smaller pods are attached to the rear fuselage which house engine monitoring instruments or cameras. This aircraft ('760') is operated by the CFTE and has been involved in various engine tests. Since April 2013 one WS-20 turbofan has been fitted.

The future

Due to the limited number of Il-76MD/TDs available, further conversions have ended and in recent years the PLAAF have concentrated on the KJ-500. Following the latest reports, XAC is already working on a successor called the KJ-3000 and based on the Y-20A. The AESA radar will also be developed by the No. 14 Institute.

Special Mission aircraft currently in use with the PLAAF

Type	Alternative designation	Role	Service entry	No (est.)
737-300CP		Command post, missile tracking	2000	2
Learjet 35/36		ELINT	2003?	6
Tu-154M/D		ECM/ELINT	1992	9
Y-8CB	Y-8GX-1	ELINT	2003?	5
Y-8G	Y-8GX-3	ECM/ELINT	2005?	7
Y-8T	Y-8GX-4	Command Post	2005?	5 or 6
Y-8W (KJ-200)	Y-8GX-5	AEW	2007?	5
Y-8XZ	Y-8GX-7	Psy Warfare	2008/09	2 or 4
Y-9XZ	Y-8GX-9	ELINT	2014	4?
Y-9W (KJ-500)	Y-8GX-10	AEW	2015	some 8, i/p
Y-9G	Y-8GX-11	ECM/ELINT	2017	2-4
Y-9X	Y-8GX-12	ECM/ELINT	2018	2?

Helicopters

Changhe Aircraft Industrial Corporation (CHAIC) Z-8 Super Frelon

This, for many years largest helicopter in PLA service, was designed by the No. 602 Institute/Changhe Aircraft Industrial Corporation (CHAIC) as a development of the French SA321Ja Super Frelon, 13 of which were acquired in the late 1970s. This helicopter was developed in the 1980s as a land- or ship-based medium-sized ASW/SAR helicopter and several Z-8s in different versions have been delivered to PLAN as Z-8J transports, Z-8S naval SAR versions plus another SAR variant called Z-8JH fitted with medical equipment. It has a maximum take-off weight of 13 tons, a cruising speed of 248km/h (154mph), a service ceiling of 3,050m (10,007ft) and is powered by three WZ-6 turboshafts.

As well as the naval variants, there is also an army transport variant called Z-8A. It was certified in February 1999 and delivered to the Army for evaluation in 2001 but due to reliability and performance issues only a small batch was delivered to the Army in November 2002. The first variant to enter wider service became the much improved PLAAF version Z-8K powered by upgraded WZ-6G turboshafts and equipped with a glass cockpit. Several dozen have been delivered since 2007 as Z-8K SARs and Z-8KA CSARs to rescue downed pilots or paratroopers. In contrast to the early Army variants they feature a FLIR turret and a searchlight underneath the cabin, RWR antennas on both sides of the nose, a hoist, and flare dispensers attached to the fuselage. Several were fitted with a terrain-following radar under the nose and a few have a sand filter installed on top of the engine intakes. The final variant is the Z-8KH which is also operated by the PLAAF in Hong Kong, and which features additional chaff and flare dispensers installed in the floats. Besides these versions, the PLA Army has introduced a similar variant Z-8B since 2011 with the floats removed to reduce weight. As in Naval Aviation, they are expected to be replaced by the latest Z-8G in PLAAF service.

So far the heaviest helicopter in Airborne Forces' service, the Z-8KA is used for CSAR, transport of special forces and regular parajump.
(bassman via FYJS Forum)

Harbin Aircraft Industrial Group (HAIG) Z-9 Haitun

The current standard helicopter is the Z-9 series, itself a licenced development of the original AS565N Dauphin, which was initially a licence-built variant known as Z-9A/B. Later, an uprated version based on the AS565SA Panther, was developed by the No. 602 Institute and Harbin Aircraft Industrial Group (HAIG).

The initial Z-9A light-transport variant was not used by the PLAAF or Naval Aviation and the first to enter PLAAF service were a few Z-9Bs procured for the PLAAF special forces unit in Hong Kong ('600x', '610x'). In 2010 another batch of new Z-9s entered service with PLAAF and featured the revised bigger nose cone of the Army Aviation Z-9W. This type is a dedicated anti-armour variant and was at first equipped with four KD-8 wire-guided anti-tank missiles. Besides the ATGMs, it can carry two 57mm (2.24in) or 90mm (3.54in) rocket pods, two 12.7mm (0.5in) machine gun pods, two 23mm cannon, or four TY-90 IR-guided AAMs. The main sensor is a roof-mounted optical sight. In PLAAF service this variant was superseded by the improved night-attack version, Z-9WA which, since 2000, featured more powerful engines and a pair of

stub wings which are able to carry up to KD-8 ATGMs or PL-90AAMs. The main sensor
on the improved variant is a low-light TV/IRST turret relocated from the roof to the
redesigned nose. The final improved version is the Z-9WZ, which is powered by two
uprated WZ-8H turboshafts and features an improved fire-control system including a
laser designator which allows the helicopter to fire the new KD-9 ATGM. It entered
service with Army Aviation brigades in early 2005 and since 2007 the PLAAF has taken
delivery of a few examples. Again, a modified variant Z-9ZH is in service in Hong Kong.

Changhe Aircraft Industrial Corporation (CHAIC) Z-10 Thunderbolt

For many years the PLA Army Aviation lacked a true attack helicopter and, despite the
introduction as an interim solution of several combat-capable Z-9Ws, a modern combat type was urgently needed. Development began as the civil-covered China Medium
Helicopter (CHM) programme in 1994 by both the Nos 602 and 608 Research Institutes
in order to gain access to Western cooperation. Technical assistance was received
from AgustaWestland (transmission), Eurocopter (rotor design) and Pratt & Whitney
Canada (PT6C-76C engine). In addition, a secret contract was signed with the Russian Kamov design bureau to design and verify the airframe and propulsion system.
A preliminary concept was presented in 1995 and, following several reconsiderations,
full development began in 1998 at the No. 602 Institute. By then the main development
team had changed from HAMC to CHAIC. First ground tests of the rotor system and
transmission took place in 2002 and the first Z-10 prototype, which had evolved into
a type comparable to the Italian A129, first flew on 29 April 2003. Altogether eight

prototypes were built before the flights were concluded in 2008-9. During that time several modifications and changes were introduced, particularly those related to the powerplant. Although the prototypes used the Canadian engine and were evaluated by the Army Aviation in 2007, the serial production variant needed a new engine due to the embargo. Therefore a 'weight-reduced' version was developed in 2009 powered by the less powerful WZ-9 engines (1,000kW/1,341hp) and, having eliminated certain non-critical parts, structures and systems, certification was granted in October 2010, with the first batch of pre-serial Z-10s entering Army Aviation service in late 2010.

The Z-10 follows the standard gunship helicopter configuration with a narrow fuselage and stepped tandem cockpit with the gunner in the front seat and the pilot behind, but unusually, particularly for a Kamov design, the fuselage has a stealthy diamond-shaped cross section. The rotor system consists of a five-blade main rotor made of composite materials and an X-style four-blade tail rotor. The main weapons are eight KD-9 or KD-10 ATGMs, a 23mm PX-10A chain gun and rocket pods. Other options include PL-90 AAMs, various rocket pods and external fuel tanks. The main sensor is a nose-mounted PNVS and TADS housing FLIR, TV camera, laser range finder and designator as well as RWR and pulse-Doppler radar, MAWS antennas and LHRGK003A laser warning receivers. In addition, the Z-10 is fitted with an integrated communication/navigation system, a comprehensive ECM suite, IFF, chaff and flare dispensers, 1553B data bus, HOTAS and a modern glass cockpit.

Three variants are so far in service with the PLA: a few Z-10Hs, which are believed to be from the initial LRIP-batch since they are equipped with the original PNVS/TVDS system and feature the original airframe configuration with a larger tail fin, the standard Z-10A Army Aviation serial variant and, since July 2015, an improved variant Z-10K.

First noted in 2015, the Z-10K is an improved variant of the Z-10A. It features an enhanced targeting system with an additional sensor behind the PVDS, a new 23mm gun and allegedly upgraded engines. (bassman via FYJS Forum)

This last example features an improved targeting system with an additional sensor behind the PVDS, a modified or new 23mm gun and, allegedly, upgraded engines. In contrast to the Army version, these wear a three-tone camouflage and have also been seen carrying a larger 19-tube 70mm (2.75in) rocket launcher similar to the American M261 for a greater firepower against ground targets.

The future

Since March 2014 Army Aviation Z-10s have been practising amphibious operations including sea-based exercises and it is therefore thought that the Z-10 could be introduced into the PLAN Marine Corps but it remains to be seen if more Z-10Ks will be procured by the PLAAF. In the near future the Z-10 is expected to be powered by the WZ-16 turboshaft engine (about 1,500kW/2,012hp), which will no longer limit is flight performance and weapons-carrying capabilities.

Future types

For the Airborne Forces, two helicopter types, which are have either been recently introduced or are in the final stages of development are noteworthy. The first is the dedicated transport variant of the naval Z-18, which was originally known as the Z-18A but entered service in early 2018 as the Z-8G. The Z-18 is the military variant of the civil Avicopter (AVIC Helicopter Company) AC313 – also known as the Changhe Z-8F-100 – based on an updated design of the earlier Z-8. It is intended as a modern 13 to 14t helicopter to replace the Z-8. It uses composite materials for the rotor blades, has a titanium main rotor head and is equipped with a modern integrated digital avionics system and an advanced electronic flight instrument landing system. Most significantly however, it features a redesigned fuselage with a larger internal volume and is fitted with improved WZ-6C turboshafts (about 1,300kW/1,743hp) giving better performance at higher altitudes and temperatures. The first prototype first flew at Jingdezhen on 18 March 2010 and is designed to carry 27 passengers or a payload of five tons and has a maximum range of 900km (559 miles).

In contrast to its naval derivates it features a further modified fuselage with a repro-filed nose section similar to that of the Mi-171 and S-92. It can accommodate 27 fully armed troops or 15 medical stretchers, has a range of 800 to 1,000km (497 to 621 miles) and is especially tailored to mountainous operations up to 8,000m (26,247ft) in the Tibetan plateaus. It is more powerful and better suited to this than the earlier Z-8B model or the Russian Mi-171 helicopters. An AEW variant (Z-18J) and an ASW/anti-ship variant (Z-18F) have already been deployed in Naval Aviation service and a transport version and other additional variants have been developed or are still under development for all PLA branches.

Following the latest images, this new version – similar to the smaller Z-20 – features a terrain-following radar mounted under the nose and it has a SATCOM dome as well as Beidou/GPS antennas installed on top of the tail boom for long-range communications. The second type will be a future medium-weight helicopter intended to replace the Z-9 series and will most likely be a variant of the Z-20. This type of helicopter in the 10-ton class has been under development as the general purpose 'China Medium

The most likely successor to the Mi-171 will be the new Z-20, which is being introduced within Army Aviation units. It reportedly offers superior performance especially in hot and high environments. (haohanfw Forum)

First known as the Z-18A, this type entered service in Army Aviation Brigades in 2017 as the Z-8G. It would be an ideal successor to the Z-8KA. (Feng Jian via CDF)

Helicopter' (CMH) at the No. 602 Institute, Harbin and Changhe based on the American S-70C design. The Z-20, however, features a redesigned cabin of slightly increased volume and a five-blade main rotor. The first prototype made its first flight on 23 December 2013 and currently there are several prototypes under OPEVAL test for the Army Aviation and additional variants for naval roles are reportedly planned. Another option might be a version of the AVIC AC352 – with the possible PLA designation Z-15 – which is a seven-ton class medium-weight utility helicopter jointly developed with Airbus Helicopters and HAIG as the H175. The AC352 made its first flight on 20 December 2016.

European and Russian helicopters

Eurocopter EC225LP/AS332L1 Super Puma

Acting as dedicated VIP helicopters, six Aerospatiale AS332L1s were acquired from France in 1986, and are rarely seen. Similarly, a second batch of three new Eurocopter EC225LPs was purchased in 2011 and little is known about their fit. Based on images that appeared in August 2017 it seems as if the Super Pumas have been modified with an additional Beidou navigational antenna installed on top of the tail boom.

Mil Mi-17/Mi-171

(ASCC 'Hip')

Rarely photographed, the PLAAF uses six AS332L1s – as seen here – and three EC225LP helicopters for VIP transport. (CMA)

Probably the most important tactical transport helicopter in PLA service is the Mi-17 or Mi-171 of which several variants have been in service, notably with the PLA Army Aviation, since 1991. A description of all subtypes, variants and developments will be listed in the Army Aviation book due for publication in April 2019, and only those types in use by the PLAAF will be covered here. The main types in PLA service are the regular Mi-17, which was introduced in 1991, followed by the improved Mi-171 which was purchased in 1995 and both have been reportedly upgraded to the Mi-171 standard in recent years. The next variant came in 2001 and this was the Mi-171V-5 featuring a solid nose, TV3-117VM engines and a hydraulically operated loading ramp. Several improved

The standard Mi-171, as here, is a cheap, robust and versatile helicopter and the backbone of the TC-subordinated transport brigades. There are also a few Mi-17V-7 and Mi-171E variants in service, recognisable by the revised front section. (DS via CDF)

Mi-17V-7s with the more powerful VK-2500 engines followed in 2003-4. These two are now the main variants flown by the PLAAF in dedicated SAR/transport brigades assigned to each Theater Command. In contrast to their Army equivalents they have been upgraded with a SATCOM antenna on top of the tail boom and a terrain-following radar mounted in a small pod under the nose. The final version which appeared 2006 is the redesigned Mi-171E used mainly by the Army Aviation. This type has a similar configuration featuring the solid nose and loading ramp akin to the Mi-17V-5. Another batch of Mi-171Es was ordered in August 2012 powered by VK-2500-03 engines and some were also delivered to the PLAAF. In the longer term, the Mi-17/171 family will be replaced by the new Z-20.

Helicopters currently in use with the PLAAF

Type	Role	Service entry	No (est.)
AS332L1/EC225LP	VIP transport	1996/2011	3/6
Mi-8	Transport/trainer	?	8?; retired?
Mi-17/Mi-171	Transport	1991/1995	some 60
Z-8K/KA/KH	Transport/SAR/CSAR	2007/2009/2010?	24/12/4
Z-9/Z-9A	Transport/utility	early 2000s	some 50
Z-9WA/WZ	Attack	2007	12/48
Z-10K	Attack	2016	16

UAV/UCAV

Beihang University (BUAA) and Harbin Aviation Industrial Group (HAIG) BZK-005 Giant Eagle

For many years, the biggest and most complex UAV in PLAAF service was the BZK-005 Giant Eagle. This type was developed by the Beihang University (former Beijing University of Aeronautics and Astronautics BUAA) and Harbin Aviation Industrial Group (HAIG) in the early 2000s as a medium/high-altitude long-range reconnaissance UAV for strategic missions. It has a cruising speed of 150 to 180km/h (93 to 112 mph), a service ceiling of 8,000m (26,247ft), and a flight endurance of 40 hours. Its maximum take-off weight is up to 1,250kg (2,756lb) with a maximum payload of up to 150kg (331lb). It was first unveiled during the Zhuhai Airshow in 2006 and features a stealthy fuselage, is powered by a piston engine driving a three-blade pusher propeller and has twin tail booms with V-shaped tail fins. A prominent head bulge hides a SATCOM antenna to provide live data transmission over thousands of kilometres and for sensors it features a small turret underneath the nose, housing FLIR/CCD cameras. The BZK-005 is currently operationally at a Strategic Scouting Unit assigned to the PLA's Headquarters close to Beijing at Shahe and also at the Naval Aviation base at Daishan Island within the East China Sea close to the Zhoushan naval base reportedly named Sea Eagle.

Northwest Polytechnic University (NTU) BZK-006A

The first tactical reconnaissance UAV in widespread service within the PLA Army is the BZK-006, which is also known as the WZ-6A or K/JWR6A. As a lightweight medium-range UAV it is launched off a truck via rocket-assisted take-off (RATO), it lands with a parachute and is a development of the earlier ASN-206/207 general purpose recon-

The BZK-005 is a relatively large UAV with characteristic twin tail booms and two V-shaped tailfins. The type has been deployed at several widely dispersed locations including the Spratly Islands facing Southeast Asia as well as in Tibet, facing India. (haohanfw Forum)

A BZK-006 during its typical rocket-assisted take-off from a truck. Besides typical reconnaissance missions, there are additional variants for artillery directing, communication jamming, communication relay, decoy, ECM and radar jamming. (FYJS Forum)

naissance-platform developed during the mid-1990s. It is fitted with a retractable turret, housing FLIR/CCD cameras for day/night missions but it can also carry a small ground surveillance radar as well. Communication is facilitated via a characteristic 'mushroom' shaped antenna on top of the head section to provide real-time datalink between the UAV and the ground command and control station. The BZK-006A is powered by a four-cylinder piston engine and has an endurance of 12 hours. Besides the regular reconnaissance variant, there are also specialised versions available for artillery directing, communications jamming and relay, as a decoy or ECM and for radar jamming.

Guizhou (GAAC) and BUAA BZK-007 Sunshine

At first sight resembling a small unmanned sports plane, the BZK-007 is complementing the larger BZW-005 but it is also rarely photographed. (CJDBY Forum)

In PLAAF service, and operational alongside the larger BZK-005, the BZK-007 acts as a medium-altitude/long-endurance (MALE) UAV and was co-developed by GAAC and BUAA in the early 2000s. The BZK-007 has a length of 7.7m (25.3ft), a height of 2.74m (9.0ft), and a wingspan of 14.6m (47.9ft). Its maximum take-off weight is 700kg (1,543lb), with a mission payload of 60 to 100kg (132 to 220lb), a maximum level speed of 230km/h (143mph), a ceiling of 7,500m (24,606ft), and has an endurance of 16 hours. This UAV – looking like an unmanned small sports aircraft – had its first flight August 2005, initially as a civilian remote sensing system called Harrier Hawk. Later it was introduced into PLA Army and Naval Aviation service. The PLAAF reportedly also uses a few, as a tactical reconnaissance UAV carrying a variety of equipment including FLIR, CCD TV cameras, as well as remote sensors of different spectral bands. This UAV is powered by a piston engine driving a three-blade propeller.

China Aerospace Science and Technology Corporation (CASC) BZK-008

Only recently noted, the small tactical reconnaissance UAV called the BZK-008, entered service within the PLA Army in 2011 but is now also in service within the Naval Aviation and PLAAF. It takes off via RATO and features a retractable EO turret containing FLIR and CCD cameras for both day and night missions.

Guizhou (GAAC) BZK-009 Sky Wing/Wind Shadow/Cloud Shadow

Still mysterious is one larger UAV, originally unveiled in 2004 as the WZ-2000 and reportedly also known as WZ-9 or even BZK-009. This UAV is a jet-powered (reportedly using a WS-11), stealthy, long-range reconnaissance UAV which has been under development at GAAC since 1999. Externally, it looks like a miniaturised and stealthier US RQ-4 Global Hawk featuring a fuselage with flat bottom and swept wings. It is powered by single turbofan engine (WS-11) with the exhaust nozzle shielded by twin V-shaped tail fins and features a prominent bulged head containing a large satellite communication antenna. It reportedly has a length of 7.5m (24.6ft), a wingspan of 9.8m (32.1ft), a maximum take-off weight of 1.7 tons, with a mission payload of 80kg (176lb), a maximum level speed of 800km/h (497mph), a ceiling of 18,000m (59,055ft), a combat radius of 800km (497 miles) and a flight endurance of three hours. It is said to carry FLIR and CCD cameras inside a turret underneath its nose for navigation and reconnaissance and possibly a synthetic aperture radar (SAR) could be installed within its fuselage.

Following its appearance at Zhuhai in 2004 nothing more was heard about it, although several reports state that the original WZ-9 first flew in December 2003, the improved BZK-009 flew first in 2006 and the type has been in limited service within the PLA Department of General Staff since in 2007 for strategic reconnaissance missions. However, not a single image of the aircraft is known.

It is known, however, that in October 2008 a similar UAV was seen at the CAC airfield in Chengdu. This type is known as Sky Wing and it later reappeared at Chengdu in 2012 and again in revised form in 2014. At first designated Sky Wing II, it soon became public that this – now twin-engined UAV (WS-500) – is already the Sky Wing III, which was later renamed Wind Shadow. During the Zhuhai Airshow 2016 yet another revised version was unveiled and offered for export, this time powered by only one turbofan but designated Cloud Shadow. All three were developed by the Chengdu Aircraft Industry Group (CAIG) in cooperation with Guizhou Aircraft Industry Corporation (GAIC). As such, it seems as if the original WZ-2000/-9-design did not materialise, and that WZ-9/BZK-009 and Sky Wing are the same project with the Sky Wing I being the very first prototype, which later evolved into two separate projects, namely the single-engined Cloud Shadow and the twin-engined Wind Shadow. In its current form, the design of both Shadow-drones evolved more like a General Atomics Avenger-comparable system than a mini-Global Hawk.

The Cloud or Wind Shadow UAV remains a mystery. Officially, a type called BZK-009 is in service, but no image is known. On the other hand there are several types – known as the Sky Wing, Wind Shadow or Cloud Shadow – under test but no confirmation is available. (CCTV via CDF)

No. 611 Institute and Guizhou (GAAC) Soaring Dragon II

So far known as the Soaring Dragon also known as the EA-03, this long-range semi-stealthy UAV is also being developed by the No. 611 Institute/CAIG and GAAC. It is unique and has no equivalent in the West. In comparison to the former family of UAVs this system is substantially larger with a length of about 14m (45.9ft), a wingspan of more than 22m (72.2ft) and a height of more than 4.5m (14.8ft). It weighs 7,500kg (16,535lb), has a speed of 750km/h (466mph), a range of 7,000km (4,350 miles), an endurance of 10 hours and service ceiling of 18,000m (59,055ft). The unique feature, however, is its box- or diamond-wing design to increase lift while at the same time reducing drag and weight. First noted in model form in 2009 and then seen at Chengdu in June 2011, the EA-03 was built at first as a technology demonstrator Soaring Dragon I for ground-based and RCS testing. It was repeatedly seen at the CAC facility undergoing tests but it never flew. It subsequently underwent a substantial redesign based on issues revealed during the tests and in that form the new Soaring Dragon II or EA-03 appeared in November 2013 featuring several major changes: it is smaller in size and now has two vertical slanted tail fins which extend outwards. It also has two ventral stabilising fins and has a much lower profile than its predecessor. Its engine is reportedly a domestic version of the AI-222-25 turbofan without reheat developed by the No. 649 Institute. Besides the prototype seen in 2013 little is known and a reputed first flight in late 2012 at the GAAC airfield is unconfirmed. The Soaring Dragon II has been in serial production since 2015 at GAAC and has entered service with both the PLAAF and the Naval Aviation as a HALE ELINT UAV. It is said to fly long-range reconaissance and EW missions and, reportedly, is equipped with a comprehensive reconnaissance and EW suite.

The unique Soaring Dragon UAV is probably one of the most secretive intelligence-gathering assets in operational use with the PLAAF. Images are rare and this is the only known photo showing it in flight. (CDF)

A standard WD-1K Wing Loong I. This type was the first true UCAV to enter PLAAF service and is known to be in service with at least two brigades. (Yang Gao)

No. 611 Institute/CAC WD-1K Wing Loong I and WL II Pterosaur

The first second-generation UAV to resemble US-designs is the WD-1K Wing Loong I – also known in the West as Pterosaur I – which is a medium-altitude long-endurance (MALE) UCAV similar to the US MQ-1 Predator. This UAV is 9.05m (29.7ft) long, has a span of 14m (45.9ft) and is 2.77m (9.1ft) high. It has a weight of 1,100kg (2,425lb), a maximum speed of 280km/h (174mph), a range of 4,000km (2,485 miles), flight endurance of 20 hours, a service ceiling of 5,000m (16,404ft) and a weapons load of 200kg (441lb). The WD-1K, as this type is designated in PLAAF service, has been developed by the No. 611 Institute/CAC and GAAC since 2005, making its first flight in October 2007. As with the Predator, it features a bulged head which houses a SATCOM antenna for communications and guidance with the ground control station via a satellite or via direct signal transmission. It is powered by a 100hp ROTAX 914 piston engine. Its sensors are an EO turret under the nose comprising a FLIR/TV/laser range finder/laser designator for a standard weapons load of two KD-10 laser-guided ATGMs. As an alternative it can also carry two PL-90 IR-guided AAMs. The WD-1K was 'unveiled' as operational during the SCO Peace Mission 2014 military exercise and is currently in service with the PLAAF Flight Test and Training Base and an operational brigade.

In September 2015 it was reported that CADI/AVIC was working on a larger development of the original WD-1K designated Wing Loong II. This new UAV, which was at first revealed via an information sheet and exhibited in model form at the Beijing Aero Show, is a turboprop-powered (possibly a WJ-9 456kW) UAV which is able to carry up to 12 ATGMs. Like its predecessor, the Wing Loong I, it bears a distinct similarity in appearance to the Predator/Reaper family of UAVs developed by the US and the Wing Loong II represents a similar step from the Wing Loong I in the same way as the Reaper evolved from the Predator. Following data published during Zhuhai, it is 11m (36.1ft) long, 4.1m (13.4ft) in height and a has span of 20.5m (67.3ft). The maximum flying altitude is 9,000m (29,528ft), with a flying speed of up to 340km/h (211mph). It has a maximum take-off weight of 4.2 tons, with an external carriage weight of 480kg (1,058lb), and can fly for 20 hours in a persistent mission cruise. This is well below the capability of the Reaper but nevertheless represents an impressive improvement over its predecessor.

The prototype of the much improved Wing Loong II during its maiden flight. This type is known to be in production for a foreign customer but operational PLAAF service is unconfirmed.
(CDF)

The future – not yet operational

No. 601 Institute/SAC Divine Eagle AEW

The next system has no counterpart in the West, although the US did have a similar system under consideration with its Sensorcraft project. The Divine Eagle is a very large high-altitude long-endurance (HALE) UAV which has been under development at the No. 601 Institute/SAC for the last decade as a dedicated 'anti-stealth' AEW platform. In contrast to all other UAVs this system features a unique twin-fuselage design with very tall twin vertical tail fins and an extra long almost unswept main wing extending across the rear fuselage. An additional canard-like wing connects the front sections of both fuselages which are dissimilar. That on the port side features a SATCOM antenna installed inside a prominent bulge. Although not confirmed it is expected to be powered by a medium-thrust turbofan engine without reheat (possibly a WS-12) mounted above the main wing and between the two vertical tails. The Divine Eagle has an estimated height of 6m (19.7ft), a length of 14m (45.9ft), a wingspan of around 35m (114.8ft), an endurance of up to 12 hours and a service ceiling of up to 18,000m (59,055ft).

Its current status is unclear: reportedly the current single airframe is probably a downsized technology demonstrator and was finished in the spring of 2015. The first low-speed taxi tests began in May 2015 but since then little more became known, although some reports suggested that a maiden flight was successfully completed in October 2015. The single image from July 2016 suggests that there is only one example, but at least one more airframe is currently at GAAC. Concerning its future operational use, information is sparse: following Chinese reports, the Divine Eagle will be equipped with multiple conformal radar arrays installed along the forward fuselages as well as on the leading edge of the forward canard wing acting as an aerial AEW system. Reports suggest that its main sensor will be a VHF metre-wave radar capable of detecting stealth aircraft at a relatively long range. Although this system is probably limited by low accuracy, the use of several UAVs flying in formation and controlled via datalink by an AEW&C aircraft or by ground stations could overcome this issue.

When and if the Divine Eagle enters service is unknown but if successful it could become the first airborne anti-stealth radar system that could be particularly useful in the vast airspace over the South China Sea area or in the Eastern Theater to counter US stealth types.

The Divine Eagle is still at the prototype stage. A maiden flight has been confirmed and at least two aircraft are known. It follows the same concept as the abandoned USAF Sensorcraft UAV.
(CDF)

UCAV

Chinese design institutes and manufactures are actively involved in stealth programmes broadly similar to Western systems such as the US X-45, X-47, RQ-170 and RQ-180 or the European Neuron or Taranis. The first of these UCAVs is the so-called Sharp Sword, which is under development at the No. 601 Institute/SAC and Hongdu (HAIG). First unveiled in model form in September 2011 and looking remarkably similar to the X-45 tailless flying wing featuring a dorsal air intake. The first true aircraft – probably a technology demonstrator – was seen for the first time during high speed-taxi testing in January 2013. Its most notable feature was its engine, since, in contrast to its Western counterparts, it uses a standard RD-93 or indigenous WS-13 turbofan similar to that of the JF-17. Additional information is sparse but besides a SATCOM datalink antenna located at the dorsal air intake the Sharp Sword is expected to carry at least two GPS/Beidou guided bombs or LGBs such as the LS-500J in an internal bomb bay. It appears to have an estimated length of around 10m (32.8ft) and a wingspan of around 14m (45.9ft). The first prototype made its maiden flight in November 2013 from the GAAC UAV Test Base and there were rumours in 2016 that a second improved prototype was scheduled to fly in 2016, however, so far nothing has been unveiled. There have been several reports, including a UAV contest held in 2011 to explore carrier-capable UAVs, where a subscale UAV similar to the Sharp Sword was tested, that refer to a naval application similar to the X-47B. Other reports even hint that the prototype already features foldable wings but from the few images available this cannot be confirmed.

The Sharp Sword, however, is not the only stealthy flying wing UAV/UCAV and for many years there were reports about a type called Dark Sword. This type, which was said to be an aerial-combat optimised supersonic-capable UCAV was probably only a concept under consideration and was long expected to be abandoned. However, in early 2018 a first image of a full-scale mock up was unveiled. Therefore its current status is unclear. There is also another flying wing design: this type is sometimes called WZ-3000 or CH-X, was developed by NTU, and strongly resembles the US RQ-180.

So far it is known only from one grainy image showing two of these UAVs wrapped under tarpaulins in a hangar and in model form seen during airshows. Several reports suggest it first flew in 2012.

In contrast to the subsonic Sharp Sword, the Dark Sword UCAV is a supersonic concept aimed for aerial combat. It is similar to the 'Loyal Wingman' concept pursued by the USAF. (CDF)

The Sharp Sword is only one of several stealthy flying wing designs and a revised version is reportedly already under test. (CDF)

UAV/UCAVs currently in use with the PLAAF

Type	Alternative designation	Role	Service entry	No (est.)
BZK-005	Giant Eagle	HALE	Early 2000s	84?
BZK-006	ASN-206/207/209 – WZ-6	Tactical recce	Late 1990s	15?
BZK-007	Sunshine	MALE	Late 2000s	28?
BZK-009	WZ-9/-2000 (unconfirmed)	HALE	Late 2000s	n/k, 12?
BZK-0??	Soaring Dragon, EA-03	HALE ELINT	2015-16	some 8; i/i
GJ-1 WJ-1	WD-1K Wing Loong	MALE recce MALE UCAV	n/k	60?, n/k
J-6W	B-6	Convert. J-6, UCAV	Early 2000s	some 100

PLAAF ARMAMENT, WEAPONS AND STORES

Although there have been significant developments in the field of indigenous Chinese airborne weapons since 2001 and many new systems have been developed, tested and introduced, there is still very little known about them. One reason for this is that the PRC authorities are strongly averse to publishing imagery which shows military aircraft together with their weaponry. Consequently, for some years most attention has been directed towards weapons acquired by China from abroad including those from the former Soviet Union and Russia. Furthermore, although several advanced weapons – chiefly those offered for export – have been presented at different air shows, the systems the PLA uses on its own aircraft are far less often shown. It cannot be denied, however, that the PRC has obtained the expertise and technologies necessary to manufacture and field air-launched weapons comparable in quality and capabilities to some of the latest Western and Russian products. This chapter provides a review of what is known about air-to-air, air-to-ground and anti-ship weapons, as well as various airborne electronic pods of Chinese origin which are known to be in service with the PLAAF and Naval Aviation.

Air-to-air missiles

Luoyang PiLi-5 (PL-5)

The PL-5 (K/AKK-5) is a result of continuous attempts to improve the PL-2, developed from the Soviet R-3S (ASCC AA-2 'Atoll'), which in turn was an unlicenced copy of the US-made AIM-9B Sidewinder. Development was initiated in April 1966 by the No. 607 Institute (now the China Academy of Air-to-Air Missiles) in cooperation with the No. 612 Research Institute (later reorganised as the Luoyang Electro-Optics Technology Development Centre – and now as the China Air-to-Air Guided Missile Research Institute). Production was undertaken from 1982 by the Hanzhong Nanfeng Machinery Factory (No. 202 Factory). Originally equipped with semi-active radar homing (SARH), but eventually cancelled in 1983, the IR-guided PL-5B, which was almost identical to the AIM-9G, was manufactured in only very limited numbers and was soon replaced by the PL-5C, which entered service in the mid-1980s. The current variant is the PL-5E-II, introduced in the early 1990s and is said to feature all-aspect and high off-boresight capabilities, and a laser proximity fuse similar to that of the later PL-8. Notably, the

Still operational on a few types, the PL-5E-II is the oldest AAM in PLAAF service, seen here on a J-7B from the 99th Air Brigade. (KJ.81.cn)

For many years, the PL-8 was the PLAAF's standard short-range IR-guided AAM. This is a PL-8B on a Naval Aviation JH-7A. It is slowly being phased out and replaced by the PL-10. (CDF)

PL-5E can be equipped with two types of warhead: a blast fragmentation type with an IR proximity fuse and an expanding rod warhead with a radio-frequency proximity fuse. The PL-5E has an all-aspect capability with the seeker having a maximum off-boresight angle of plus or minus 25 degrees before launching, and plus or minus 40 degrees after launching.

Luoyang PiLi-8 (PL-8)

The emergence of the PL-8 (K/AKK-8) was a result of semi-secretive Sino-Israeli cooperation during the 1980s. Following the impressive achievements of the Israeli-made Python 3 AAM in air combat over Lebanon in 1982, Beijing approached Israel with a request for technology transfer. A contract signed on 15 September 1983 resulted in Israel providing not only samples of the Python 3 but also 1,200 kits for licence production of the missile by the Xi'an Eastern Machinery Factory. The first batches of PL-8s became available to the PLAAF in early 1988. Subsequent development of the PL-8 was managed by the Luoyang Electro-Optics Technology Development Centre, which aimed to gradually reduce the Israeli-made content, resulting in the PL-8A, which entered production in the late 1990s. The current exclusively domestically-made variant PL-8B was first noted in mid-2005 and featured a PL-9-style all-aspect Indium antimonide (InSb) seeker and a programmable digital processor, which offer a wider off-boresight angle. Its range has been increased to 20km (12 miles). The missile is also compatible with Chinese made HMS and is the only variant in use. The same technology transfer seems to have included the supply of the Israeli-made Display And Sight Helmet (DASH), made by Elbit Systems, since the PLAAF subsequently introduced a similar system on its J-7IIAs. The PL-8B is currently the primary WVR weapon in PLAAF and Naval Aviation service but will replaced by the new PL-10.

The PL-10 and PL-15 entered service in early 2017, replacing the PL-8B and PL-12 combination on J-10B and J-10C fighters. (Top.81 Forum)

Luoyang PiLi-10 (PL-10)

Still surrounded by much mystery, the PL-10 (K/AKK-10) was also known as the PL-ASRM, which stands for PL-Advanced Short-Range Missile. Its protracted development was initiated in response to a PLAAF request for a missile that could match such Western weapons as the AIM-9X, AIM-132 ASRAAM, A-Darter, AAM-5 and IRIS-T. The PL-10 features a multi-element imaging infrared seeker, a laser proximity fuse and thrust-vectoring nozzle, and offers a 90-degree off-boresight capability with a maximum load of 55G, as well as excellent countermeasures discrimination. It is powered by a solid rocket motor. The latest reports indicate a lock on after launch (LOAL) capability, which could extend its range well beyond the visual arena. The PL-10 is compatible with the latest HMDs. Similar to other modern AAMs the current configuration features four control tail fins complementing four narrow stabilising strips/wings attached to the mid-section of the missile body which has a length of 2.96m (9.7ft), a diameter of 0.16m (6.3in) and weight of 105kg (231lb). Development began in 2005 at the China Air-to-Air Guided Missile Research Institute (former Luoyang Electro-Optical Centre) and the first test round was apparently fired in November 2008. The early configuration (pre-2013) seen for the first time in early 2011 on J-11B fighters, differed considerably from the current design and, allegedly, the first test launches took place in November 2008. It was first seen in 2011 and subsequently entered initial production in 2013. The PL-10 AAMs entered service with the PLAAF in late 2016 and it is expected that the Naval Aviation will follow soon.

SAST PiLi-11 (PL-11)

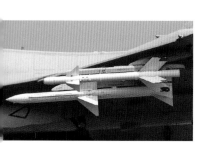

The early PL-11 as seen here on a J-10A was the PLAAF's first medium-range AAM and entered service on J-8H and early J-10A fighters. However, it was never entirely successful (VF-154 via CDF)

Following several failed attempts to reverse-engineer the US-made AIM-7 Sparrow medium-range SARH missile in the 1970s and early 1980s, China purchased the Italian-made Alenia Aspide 1 as a surface-to-air variant in the mid-1980s. Corresponding agreements for licence manufacture resulted in the HQ-61C SAM, which in turn led to an AAM variant. Developed since 1987 by the Shanghai Academy of Science and Technology (SAST) the first locally manufactured PL-11 (K/AKK-11) appeared in 1989, but subsequent cooperation with Italy was cancelled following the events at Tiananmen Square. Due to delays, the first test-firings were undertaken from a J-8B test-bed in 1992 and the weapon entered service in the mid-1990s. Final certification tests took place in 2001, by which time it was already considered obsolete. The standard PL-11 featured a mono-pulse SARH seeker, which was updated incorporating inertial guidance so that targets would need to be illuminated only at the final stage of the engagement. This weapon saw only limited service as the PL-11A in regiments equipped with J-8Ds and J-8Hs and it has probably now been replaced by the PL-12. Alleged reports about an active-radar-guided variant called PL-11B are either wrong or that variant never entered service.

LETRI PiLi-12 (PL-12)

Development of China's first true active radar-homing AAM in the PLAAF arsenal was initiated by the Leihua Electronic Technology Research Institute (LETRI) and the No. 607 Institute in the early 1990s, with the intention of developing an active radar-homing missile comparable to the AIM-120 AMRAAM. The centrepiece of this project was the development of a miniaturised active radar seeker, sometimes identified as the AMR-1, allegedly based on the Russian 9B-1348 seeker for the R-77 and its datalink, apparently in cooperation with the AGAT Research Institute in Moscow. The seeker was unveiled for the first time at the Airshow China at Zhuhai in 1996 and the resulting PL-12 (K/AKK-12) was completed in December 2004 and was certified in 2005. Compared to the AIM-120, it is slightly longer, wider and heavier, has tail-mounted control fins of slightly greater span with a distinctive notch cut into their bases. Additionally, the

Replacement for the semi-active guided PL-11, the active-guided PL-12 became the PLAAF's standard medium-range AAM in the early 2000s. Here it is carried on a J-11B. (Top.81 Forum)

In 2017 the J-16 became the second type to use the PL-10 and PL-15 in operational service. The PL-15 with its cropped fins is also the main armament for the J-20A. (DS via SDF)

PL-12 features two datalink antennas next to the nozzle for mid-course correction and several dielectric strips along the middle warhead section which house the radio prox-imity fuse. A variable-thrust solid rocket motor offers two levels of power for different sections of the flight envelope, resulting in an AAM which is considered superior to the AIM-120B, but slightly inferior to the AIM-120C. It first became operational with the J-8F and from then on has been the standard long-range AAM on all J-8F/DFs, J-10s, J-11Bs and J-15s and also, due to a radar modification of the fighter, on the Su-30MK2. The early PL-12s were to be superseded by the PL-12A, featuring an improved seeker with a new digital processor and SINS, with the result that the PL-12A is rated com-parable to the AIM-120C-4. Since 2010 there have been reports concerning additional upgrades, some claiming that he PL-12 may have an active/passive dual mode seeker in order to achieve greater ECCM capability and kill probability. In addition, several improved versions comparable to the American AIM-120D were proposed. These include a PL-12B with an improved guidance system, a PL-12C with foldable tail fins for internal carriage and a PL-12D with a ramjet motor for long-range engagements although none of these seem to have been successful, at least as PL-12 variants.

LETRI PiLi-15 (PL-15)

It is most likely that these concepts evolved into a missile which was initially speculated as the PL-12C but is now known by its new designation PL-15 (K/AKK-15). In contrast to the PL-12, it features redesigned, cropped main and tail control fins with a smaller wingspan in order to fit into the internal weapons bay of the J-20A. The new missile is powered by a new a dual-pulse rocket motor which extends its range to 200km (124 miles) and its two-way datalink plus a new active/passive dual mode AESA seeker on board enables excellent ECCM capabilities. Development of the PL-15 began in 2011 at the No. 607 Institute and a first demonstrator missile was seen in 2012 on a J-11B. Later in 2013 during weapons integration tests it was also noted in the J-20's main weapons bay. The first successful test-launches took place in 2015 and in November 2016 this missile was seen in operational service on J-10Cs and J-16s. It is thought to replace the PL-12 as the PLAAF's main long-range AAM on all modern front-line fighters.

Surprisingly, in late 2016 J-16 prototype '1603' was seen carrying a PL-X, a weapon in a class of its own in terms of range.
(CDF)

PL-X (possibly PL-16)

The final AAM currently under development is a very long-range air-to-air missile (VLRAAM) so far only designated PL-X. This missile is most likely the successful contender of the original VLRAAM-requirement, for which also the PL-12D and PL-21 with ramjet-motors were possibilities. In comparison to the PL-12/-15 it is significantly longer and features a wider body closely matching the size of a SAM. According to rumours, the PL-X is powered by a dual-pulse rocket motor in favour of a ramjet engine, which results in a more compact and slimmer design in order to achieve a long range of more than 300km (186 miles), a speed in excess of Mach 4 and a cruising altitude of 30km (19 miles). The PL-X has only four small tail fins, which are coupled with TVC and the missile is guided by an advanced two-way datalink and an active AESA seeker with enhanced ECCM capability.

Some sources assume it to have an additional IIR seeker as indicated by a small optical window in its nose, which would further increase its kill probability amid severe jamming. It is expected to fly a semi-ballistic trajectory similar to the US AIM-54 in order to achieve an extra-long range. Consequently, it cannot be carried internally but has to be mounted externally on long-range fighters/interceptors such as the J-11B/D, J-16 and the J-20. The PL-X is a new class of AAM and is slated against high-value aerial targets deep behind enemy lines such as AEW&C aircraft and tankers. Based on the latest reports, a first test was successfully accomplished in November 2016 when a PL-X was fired from a J-16.

Russian AAMs

Combined with the arms packages acquired with the Su-27SK and Su-30MKK/MK2 purchased during the 1990s the PLA introduced several Russian AAMs. An exact description of these systems – and other Russian air-to-ground weapons as well – can be found in the book *Russia's Air-launched Weapons* (Harpia Publishing, 2017). These missiles include the R-27, the R-73 and the R-77. The following paragraphs therefore include only PLAAF-related information.

Together with the Su-27/30 arms package, China acquired several AAMs from Russia including the well known R-73E (foreground) and R-27R1. Less well known are the longer-range R-27ER1 and R-27ET1, also used by J-11A fighters.
(CDF)

Vympel R-27ER1/ET1 (ASCC AA-10 'Alamo')
Originally, the PLAAF acquired several R-27R1s to be used by the Su-27SK and J-11A and only recently has it become known that a few hundred R-27ER1s were also acquired. This variant is a semi-active radar homing AAM featuring a larger and longer rocket motor in comparison to the standard R1 giving it a maximum range of 100km (62 miles) compared to the original 70km (43 miles). Less well known is that several R-27ET1 IR-guided AAMs were also purchased and the longer-range R-27ER1/ET1s are usually carried by J-11s together with the shorter-range R-77E.

Vympel R-73E/R-74E (ASCC AA-11 'Archer')
Together with the R-27, the PLAAF also acquired the R-73 for the Su-27SK and J-11 and its main advantages in comparison to Chinese contemporary MMAs were a substantial off-boresight capability and its ability to be targeted by a HMS. In PLA service this missile can be carried only by the original Su-27SK and J-11A fighters. There are

unconfimed reports that the the modernised R-74 missile was acquired together with the Su-35 purchase.

Vympel R-77E (ASCC AA-12 'Adder')

The R-77 – also known by its export designation RVV-AE – was Russia's answer to the US AIM-120 AMRAAM and China introduced this AAM as part of the Su-30MKK's weapon package that provided the PLAAF for the first time with a true medium-range fire-and-forget AAM with a range of just over 60km (37 miles). At least 400 R-77Es were acquired from Russia during the 2000s and are usually seen on the Su-30MKK, Su-30MK2 and late-production Su-27UBKs. In Naval Aviation service the R-77E is used only by the Su-30MK2 but is seems that it will be replaced in the long term by the indigenous PL-12 or even the PL-15. The latest acquisition of a Russian AAM seems to be the upgraded longer-range development of the original R-77 as part of the Su-35 purchase in late 2015 but the R-77-1 – also known as RVV-SD – has so far been rarely seen in service.

The R-77E was introduced as part of the Su-30MKK's weapon package, which provided the PLAAF for the first time with an active-radar-homing AAM, enabling the development of new tactics to engage enemy aircraft in pure BVR scenarios. These AAMs are being prepared for J-11A fighters. (Top.81 Forum)

Air-to-air missiles

Name/type	Current variant	Guidance	Range	Date of introduction	In use with
Luoyang PL-5	PL-5EII	IR	0.5–16km	First half 1990s	J-7, JH-7A
Luoyang PL-8	PL-8B	IR	20km	Since early 1988	J-7E/G, J-8H/F, J-10, J-11
Luoyang PL-10	PL-10	IR	20km	In introduction 2017–18	J-10C, J-16, J-20
SAST PL-11	PL-11A	Mono-pulse SARH/INS	40–75km	2002	Most likely with-drawn from use
LETRI PL-12	PL-12A	Active radar	60–70km	2005?	J-10, J-11B
PL-15	PL-15	Active radar	180-200km	2016	J-10C, J-16, J-20
PL-XX		Active radar; maybe add. IR	+300km	Still in test	J-16, J-20
Vympel R-27	R-27R1 R-27ER1 R-27ET1	SARH SARH IR	70km 100km 100km	1992 (in China)	J-11, Su-27SK/UBK
Vympel R-73/R-74	R-73E R-74E	IR	30km	1992 (in China)	J-11A, Su-27SK/UBK, Su-30MK, Su-35
Vympel R-77/ R-77-1	R-77E R-77-1	Active radar	80km 110km	2000+	Su-27UBK, Su-30MK, Su-35

The KD-20 seen on this H-6K is the PLAAF's first modern long-range ALCM, designed to attack a variety of fixed, high-value targets.
(Japan Ministry of Defense)

Air-to-surface missiles

CNGC KongDi-9/-10 (KD-9 and KD-10)

Both the KD-9 (K/AKD9) and KD-10 (K/AKD10) are second generation of ATGMs developed specifically for the Z-10 and Z-19 by the China North Industries Group Corporation Electro-Opticals Science & Technology Ltd (CNGC). Whereas the Z-10 can use both types, the Z-19 uses only the KD-9. Besides these helicopters in PLAAF service it is used by the WD-1K UCAV.

Both missiles use actually a common design quite similar to a US AGM-114 Hellfire albeit without the forward control fins and the KD-9 is slightly lighter and smaller than the heavier and bigger KD-10. As for the seeker, both feature a semi-active laser seeker – some sources suggest it being a derivate of the Russian Krasnopol 152mm (6.0in) laser-guided artillery projectile and it is not a fire-and-forget weapon. Images since August 2015 suggest a new version KD-10A featuring an improved laser seeker has been introduced into service and since some time there are rumours that an MMW seeker is under development.

In terms of configuration, the KD-9 and KD-10 share a similar design albeit with a different diameter body and slightly different fins. The differences are evident in this direct comparison.
(CDF)

CASIC KongDi-20 (KD-20)

The KD-20 (K/AKD20) was long known as the CJ-10 (ChangJian-10) and is an air-launched variant of the DH-10/DF-10 land-based cruise missile. Reportedly, this cruise missile is based on indigenous technology derived from several ill-fated projects pursued as early as 1977, as well as technology obtained from the Russian-made Kh-55 cruise missiles, and from studies of the wreckage of several US-made AGM-86s and RGM/UGM-109E Tomahawk Land Attack Missiles obtained in Iraq, Serbia and from

other conflicts. Unsurprisingly, its design closely resembles certain features of these missiles including a cylindrical body with retractable wings. The missile has four foldable tail fins and a concealed engine inlet. However, it lacks any significant stealth features. The KD-20K is the first Chinese-made ALCM, is powered by a fuel-efficient turbofan engine providing a range of between 1,500 and 2,000km (932 and 1242 miles) depending on the payload carried. Although rated as a strategic weapon and as such, allegedly capable of carrying both nuclear and conventional warheads, there is so far no indication that the missile is indeed nuclear-armed but it is, nevertheless, designed to attack a variety of fixed, high-value targets.

Guidance is utilised both by INS and terrain contour matching (TERCOM) guidance (supposedly coupled with GPS/Beidou) for most of its flight, with digital scene-mapping area correlation (DSMAC) used for pinpoint accuracy during the final phase. The KD-20K entered service on the PLAAF's H-6Ms in the late 2000s, the H-6K can carry up to six rounds and it is expected to be used also by the new H-20 strategic stealth bomber under development, for which a new stealthy ALCM is also planned.

Allegedly, an improved version with a longer range of 2,500km (1,553 miles) is rumoured to be under development similar to an anti-ship version sometimes known as YJ-100, but neither of these have been confirmed. However, a new variant called KD-20A was first seen in August 2017, which differs by having a high-definition imaging radar similar to that of DF-21C/D AShBMs replacing the original DSMAC optical window. Reportedly, this variant has a greater accuracy at night and in poor weather conditions as well as further improved anti-jamming capabilities.

CHETA KongDi-63 (KD-63)

The KD-63 (K/AKD63), which was initially known as the YJ-63, is the latest member of China's first operational stand-off land attack cruise missiles (LACMs). The ancestor of the KD-63 is a missile requested by the Naval Aviation for an AShMs for the H-6D. The resulting missile was the YJ-6, which was certified in 1984 and entered Naval Aviation service two years later. However, in Naval Aviation service the H-6D and YJ-6 bomber/missile combination has been replaced by the more advanced H-6G and YJ-83K, and is no longer in operational service. However, the YJ-6 formed the basis of a missile also known as the HY-4/XW-41 AshMs which was powered by an FW-41B turbojet fed through a belly intake between the four tail fins. It was first tested in 2000 and features INS midcourse and terminal man-in-the-loop guidance via a small CCD camera installed at the tip of the head section plus a small UHF/VHF TV antenna on top of the

The KD-63 is the PLAAF's first standoff LACM and is based on the earlier YJ-6 AshM family. The standard KD-63/A variant (left) has a TV seeker, whereas the improved KD-63B (right) is guided by an IIR seeker.
(Left: CDF
right: CCTV-7)

At first sight comparable to the YJ-83K, the KD-88 is in fact a different missile and is used by the PLAAF as a second-generation LACM. It too is available with different seeker heads including the standard TV seeker (above) and an IIR seeker (right) on the KD-88A. (Both via CDF)

head section. It entered service as the YJ-63 or KD-63 during 2004-5 and is primarily carried in pairs by PLAAF H-6H bombers but can also be carried by the H-6K. Following some reports, the original KD-63s or KD-63As were superseded by an improved version designated KD-63B, which features an IIR seeker in place of the TV seeker, thus enabling it to be used in all weather conditions and having a fire-and-forget capability. It can be identified externally by a new conformal datalink or GPS/Beidou antenna.

CHAIG KongDi-88 (KD-88)

Although comparable in size, configuration and capabilities, the KD-88 (K/AKD88) is not a true member of the YJ-8 family; in fact, the relationship is somewhat mysterious. Development was begun in the early 2000s by the Hongdu Aviation Industry Group (HAIG) as China's first Standoff Land Attack Missile for precision strikes against high value targets. It is comparable to the US-made AGM-84E and is intended primarily to equip the JH-7A fighter-bomber. It is imilar to the YJ-83K and is also powered by a small turbojet engine giving a range of more than 200km (124 miles) and is available with different seekers. Those identified so far are the original CCD/TV-seeker (KD-88) and an IIR-seeker (KD-88A) but anti-radiation and MMW seekers are rumoured to be under development. Both variants use INS guidance, but unlike to the YJ-83K have a different datalink for the final phase of the flight provided by four small datalink antennas extending from the tips of mid-body fins and usually provided via a guidance pod (CM-802AKG) similar to that of the Russian APK-9E for man-in-the-loop terminal corrections. For some time, the KD-88 was used only by the PLAAF but images from 2016 indicate that the KD-88 can also be carried by PLAN JH-7As. In July 2018 both a J-10C and J-16 were noted for the first time carrying this missile.

Currently the sole ARM in PLAAF service, the YJ-91 is a direct development of the Russian Kh-31. (acer31 via CDF)

HAIG YingJi-91 (YJ-91)

The YJ-91 (KR-1/H/AKJ91) is an indigenous development of the Russian-made Kh-31P anti-radar-missile. Apparently fielded in two variants, and in contrast to the Kh-31P, the YJ-91 reportedly uses a single seeker capable of covering multiple frequency bands and has an increased range of up to 120km (75 miles). Currently it is the main anti-radiation missile for SEAD missions. The second variant YJ-91A is understood to be an anti-ship missile with a sea-skimming capability (which the Kh-31A lacks), and is programmed to fly 'pop-up and dive' terminal attack manoeuvres, but its present status is unclear. Because of this latter function, the anti-ship variant probably has a much shorter range, estimated to be around 50km (31 miles). As an alternative to the YJ-91, Naval Aviation also uses the Russian original Kh-31P, which was acquired as part of the Su-30MKK's weapons package. While some sources cite a production run based on 200 missiles obtained directly from Russia, others report that the Kh-31 was not acquired from Russia but from various other sources. Whatever their origins, the YJ-91/Kh-31 variants are carried by the JH-7A, J-8G(?), J-10B/C, J-15, Su-30MKK/MK2 and H-6G. Reportedly a next generation ARM is already under development.

Russian AGMs/ASMs

As an alternative to indigenous weapons, the PLAAF and Naval Aviation also use several Russian AGMs/ASMs and stores, which were introduced at the time of the Su-30MKK and Su-30MK2 purchase. These are not only the aforementioned Kh-31Ps, but also firstly the Kh-29ME, which is similar but heavier to the US AGM-65B Maverick, a TV-guided short-range AGM which entered service with PLAAF in 2002 as part of the Su-30MKK weapon's package. The second missile is the larger and even heavier Kh-59ME, which is also a TV-guided stand-off ASM carried by the Su-30MKK. A full description of Russian stores and weapons can be found in the aforementioned book, *Russia's Air-launched Weapons*.

The Russian Kh-29ME was one of the first AGMs to enter service as part of the arms package acquired together with the Su-30MKK. (zhang81zhang via FYJS Forum)

Air-to-surface missiles

Name/type	Current variant	Guidance	Range	Date of introduction	In use with
KD-9/KD-10	KD-9/KD-10/-10A	Semi-active laser	2–7km	2016	Z-10K, WD-1K
KD-20	KD-20 KD-20A	INS/TERCOM Radar	1,500–2,000km	Late 2000s	H-6M/K
KD-63	KD-63A KD-63B	TV IIR	180km	2004/05	H-6H/K
KD-88	KD-88 KD-88A	TV IIR	200km	2006?	JH-7A, J-10C, J-16
Kh-31/ YJ-91	Kh-31P/YJ-91 ARM Kh-31A/YJ-91A AShM	L117E passive seeker Radar	110km 25–70km	2004? ?	JH-7A, J-10B/C, Su-30MKK
Kh-29TE	AS-18	TV	30km	2000	Su-30MKK
Kh-59ME	AS-18	TV	115km	2000	Su-30MKK

Guided bombs

For many years until the mid-1990s the Chinese air arms lacked any kind of precision guided munitions. The first weapons able for precision-strikes were a range of Russian-made guided bombs, primarily the TV-guided 500kg (1,102lb) KAB-500Kr and 1,500kg (3,307lb) KAB-1500Kr but no laser-guided variants were purchased from Russia. Surprisingly, these systems have been seen so far only on PLAAF Su-30MKKs and not on their naval MK2s. The availability of these weapons, and the monitoring of US operations over Iraq and the former Yugoslavia, subsequently resulted in fre-

Several KAB-500Kr (seen here) and KAB-1500Kr bombs were introduced to complement the TV-guided Kh-29 and Kh-59ME. This Su-30MKK also carries a KG600 ECM pod.
(Yang Pan via chinamil.com)

netic developments in this field, which produced a plethora of different, but frequently competing, systems. In PLA service, most of the LS- series of laser-guided bombs are broadly comparable to the US GBU-16 Paveway II series but more modern systems are currently being developed and fielded.

LS-500J

The first indigenous laser-guided bomb was the LS-500J, which is very similar to the Russian KAB-500L. Although several systems have been tested since the 1980s, the LS-500J entered service only in 2003. Following reports, this LGB is guided by two parallel laser beams and flies an S-shaped trajectory as it approaches its target. However, this mechanism has a relatively low accuracy and is susceptible to jamming and poor weather conditions. As target designator pods, the LS-500J is currently guided by three different types. For the future, a modernised variant so far known only as the GB500 (or GB1) by its export designation is destined for introduction as a second-generation LGB. In contrast to its predecessor, it will feature a proportional navigation seeker providing greater precision, higher sensitivity and will be less susceptible to a windy environment particularly with regard to slow-moving targets but, except for these features, it shares the same design with LS-500J. In addition, in November 2014, a smaller variant of 250kg (551lb) and designated GB250 or maybe LS-250J was seen for the first time and tested, and reports from late 2017 suggest that it has entered full-scale production. An even smaller version, the GB100 or LS-100J, of about 100kg (220lb) has been under development since 2012. In appearance, it is similar to the French AASM weapon, featuring four large tail fins and another set of smaller forward fins. Some reports suggest it might also feature a small solid rocket motor.

Only TV-guided bombs and AGMs were acquired from Russia, so the first laser-guided bomb in PLAAF service became the LS-500J as seen on this J-10A.
(Top.81 Forum)

In parallel to these systems there are three families of guided weapons which are either under development or are available for export and known most often only by their export designations. The fact, however, that they are offered for export strongly suggests that the PLAAF and Naval Aviation have comparable systems already in service, even if these are unconfirmed. They are the LeiTeng (LT) series of laser-guided bombs of which the LS-500J is one. Others are the FeiTeng (FT) series of GPS-guided bombs resembling the US-made Joint Direct Attack Munition (JDAM) which differ in wing configuration and weight and also the LeiShi (LS) series of GPS-guided weapons which are also available in different weights, with different seekers and wings. This family is even more variable and ranges from 'small diameter bombs' to large-calibre gliding bombs or canisters similar to the American Joint Standoff Weapon (JSOW) with propulsion systems providing ranges of up to 300km (186 miles).

Russian guided bombs

Similar to AGMs, ASMs and AShMs, the PLAAF operates a few Russian systems as alternatives to indigenous weapons, which were introduced together with the Su-30MKK and Su-30MK2 purchase. The most important of these are the KAB-500KR TV-guided bombs used as 'bunker busters' against fixed, 'hard' targets. Together with the bombs themselves, a training round without the warhead was also acquired. The second missile is the larger and heavier KAB-1500KR, which is also a TV-guided bomb solely carried by the Su-30MKK. Again, a description can be found in the aforementioned book, *Russia's Air-launched Weapons*.

Guided bombs

Name/type	Weight	Guidance	Range	Date of introduction	In use with
LS-500J	500kg	laser	>10km	post 2004	JH-7A, J-10, J-16?
LS-500J mod. (GB500/GB1)	500kg	proportional navigation	?	2016–17	JH-7A, J-10, J-16?
KAB-500KR	500kg	TV	15–17km	2000?	Su-30MKK
KAB-1500KR	1,500kg	TV		2000?	Su-30MKK

ECM pods were introduced comparatively late and initially only in limited numbers. This changed in recent years with the introduction of several new systems like this KG800. (DS via SDF)

Targeting, navigation and electronic warfare pods

Although China began the development of targeting pods together with the development of LGBs during the 1980s, most projects resulted in failure. Following a protracted development by the No. 613 Institute, the first Chinese-made targeting pod closely resembled the US-made AN/ASQ-23 Pave Spike system and was equipped with an electro-optical laser designator. This pod was tested on the Q-5E/F prototypes while the latest operational Q-5L is also equipped with a laser spot tracker similar to the US-made AN/AAS-35 Pave Penny. Currently three types of targeting pod are known to have been in service since about 2006 and they seem to be specialised variants of the same type

Available in three slightly different variants, the first true targeting pod in PLAAF service is the K/JDC-10A pod as seen on a J-10A.
(CDF)

but modified for specific aircraft: the K/JDC01 (for the JH-7A), the K/PZS01H (for the Q-5L), which is not in Naval Aviation use, and the K/JDC01A (for the J-10A).

A true combined navigation/targeting pod comparable to the US LANTIRN-system is, despite the known development of a system called Blue Sky by the No. 607 Institute, not operational, but a new pod broadly similar to the US Sniper was recently seen on a J-16. Concerning the frequently-mentioned use of a locally-developed variant of the Russian-made Sapsan/Sapsan-E targeting pod nothing has been confirmed – and in fact this seems unlikely – but Chinese fighters have been seen regularly with the Russian-made APK-9 datalink pod, used to guide the Russian-made AGMs. A similar indigenous pod most often seen in conjunction with the KD-88 and YJ-91 is known as the CM-802AKG. This is carried by the JH-7A, J-10C, H-6G and H-6K.

Electronic warfare pods have become an increasingly regular sight on several Chinese aircraft for some years and the first system known is the KG-300G, developed by

Providing target information for the KD-88 AGM and YJ-91 ARM, several aircraft are able to carry a dedicated guidance pod, known as the CM-802AKG.
(Richard Yip via SDF)

Not much is known about the BM/KZ900 SIGINT/ELINT pod, which is only known to be operational on J-8H fighters. (Top.81 Forum)

the China Electronic Technology Corporation. This pod is used to jam airborne radars working in the I- and J-band frequencies and is carried by the J-8H. Outwardly similar in appearance to the KG-300G is the BM/KZ-900 SIGINT/ELINT reconnaissance pod, designed by the Southwest Institute of Electronic Equipment, and which has been in service since the late 1990s. Intended to collect radar signals over a wide band of frequencies it is integrated in the Y-8GX series and carried by the J-8H. As yet unidentified, a more modern ECM pod has been seen on naval JH-7s, JH-7As and H-6Gs since April 2008 and at least three different configurations have been identified to date, but neither a designation nor specifications are known. Each differs slightly in shape and by their antenna arrays and, supposedly, they use different frequencies and EM-spectrums. The most recent family of ECM pods are the KL600A, KL700A (to be used by all tactical aircraft) and the KG800 (on JH-7As only). These are known to differ in their sizes and capabilities, but little else is currently known of them.

The latest and now standard ECM pod for most tactical aircraft is the KG600 and KG700 family as seen here on a J-10B. (Richard Yip via SDF)

General-purpose bombs, mines, incendiary and fuel-air explosive bombs, cluster bombs and runway-penetration weapons

For some time, and particularly since both the PLAAF and the Naval Aviation have played only a minor role in providing close air support for its ground forces on the battlefield, the number of attack formations has remained limited. Their main weapons were mainly domestic variants of Russian-made FAB series general-purpose (GP) bombs of the M54 design. But by the early 1990s a series of more streamlined, low-drag GP bombs had emerged. The most frequently seen of these is the 250kg (1,212lb) Type 250-3, designed for external carriage by modern high-speed jets, which is broadly similar to the US-made Mk 82. Similar larger-calibre weapons followed as well as a number of specialised types, including the Type 250-4, which is a retarded variant for deployment from low altitudes.

For incendiary and fuel-air explosive (FAE) bombs, the PLAAF introduced several systems of the Type 250-1 to 250-3 series during the 1990s with at least five different types known. Outwardly they are broadly similar to the US-made BLU-10 and BLU-27 canisters or the MXU-648 baggage pod but none of them have been seen recently. The same can be said for cluster bombs, of which several different types were developed and entered PLAAF service in the mid-1980s also within the Type 250-1 to 250-3 series. Again, none have been seen recently, which also counts for a Matra Durandal-like runway-penetration bomb called Type 200A or 200-4 during the mid-1980s. But such systems are doubtless in service.

Bomb-loaded aircraft are rarely shown and even less is known about the different types of bombs. The best known are members of the Type 250-1 or –3 family, as seen on this JH-7A. (FYJS Forum)

It was long reported that the latest H-6K had its bomb bay removed and replaced by fuel tanks. However, it retained the standard bay and often still uses former Soviet general-purpose high-drag bombs like these FAB-500M54s. The characteristic ballistic ring on the nose acts as a vortex generator to stabilise the bomb. (Yang Pan via chinamil.com)

Nuclear weapons

Concerning the situation of nuclear weapons even less is known. Even today the only Chinese-made nuclear bombs about which anything meaningful is known are the A29-23 and the H639-23. The A29-23 was used during China's first nuclear test in October 1964 and the first thermonuclear bomb H639-23 was test-dropped from an H-6 bomber in June 1967, both over the Lop Nor testing facility. A much smaller and more streamlined thermonuclear device, designated H524-23, was test-dropped from a Q-5 fighter-bomber in January 1972, and another nuclear device designated KB-1 probably entered service with the Q-5A. All older models of nuclear weapons are understood to have been withdrawn from service, but there is no information and even less imagery available on current operational systems.

AIR FORCE TRAINING SYLLABUS

AIR FORCE TRAINING

As noted in the Naval Aviation book, undue emphasis is frequently given to the technical modernisation of the hardware, since it is at first sight more spectacular and also much easier to follow. However, the latest developments in tactics and training are probably even more important for the future outcome of any potential operational use and as one follows the frequent reports in the Chinese media, the PLA shows increased confidence in its capabilities and training conditions. Whereas not much is known of the training of naval aviators, the PLAAF is much more open in regard to its training and flight school programmes. Obviously, the PLAAF is following a system more akin to Western concepts. In order to gain experience, not only was the training system modernised and new types introduced, but the PLAAF has been increasing its training with other air forces in recent years and it seems that the frequency, complexity and intensity of the exercises are increasing.

Therefore, this chapter provides an overview of the PLAAF's training syllabuses and describes the structure and path a PLAAF aviator has to go before he or – unlike the Naval Aviation – she earns his or her wings.

General background

Training guidance for all PLA branches still follows a strictly centralised, top-down approach with all regularities issued by the General Staff Department (GSD). These directives are increasingly focusing on 'closely realistic combat training' under all complex and adverse conditions. Concerning numbers, the PLAAF is more open than Naval aviation and official sources say that between 2013 and 2015 the PLAAF recruited about 1,000 to 1,300 new cadets each year of which a failure rate of about 50 to 60 per cent is estimated. The PLAAF has recruited aviation cadets selectively from specific provinces and municipalities based only on the fact that in rural areas only a ninth-grade education is required. However, the long-held preference of Han Chinese – due to educational and political reliability reasons – has been relaxed so that at least a small number of cadets from minorities in Inner Mongolia, Qinghai, Xinjiang and Yunnan – but still not from Tibet – are recruited.

Historically, PLAAF cadets were selected only from high school graduates and outstanding enlisted members, but in recent years this has changed so that also civilian and military college members as well as university students and graduates can enlist.

Complementing the first phases of flight training at the Aviation University and the Flight Colleges, plenty of training is performed on simple simulators in classrooms.
(CDF Forum)

Another difference from the Naval Aviation is that the PLAAF has had female aviators in an operational unit since 2012. At first, they were restricted to serve in the transport divisions, but, despite their numbers being small, their role has been expanded so that the PLAAF is now not only one of the 16 countries with female air force pilots, but they are also flying combat types including the JH-7A, J-10 and even as members of the PLAAF's '81' Aerobatic Team. However, it is not clear if male and female members are flying in mixed crews.

One additional speciality of the PLAAF's flight training is its division into a strict institutional annual training cycle, due to the PLAAF's two-year enlistment terms of conscripts, who still make up a large proportion of PLAAF service members in maintenance and other ground-support positions. This is due to adherance to the top-down training plans and PLAAF Party Committee directives, which stipulate different types of coordinated evaluations and major exercises to test the ability of pilots to perform under standardised criteria only during certain times of the year.

Consequently, the year itself is structured along five partially overlapping segments during the course of the year: 'new year flight training' (January to February), training in 'subjects and topics' (March to May), 'peak drills, exercises and evaluation' (June to August), a second round of training in 'subjects' and 'topics' (September to November) and the final 'year-end evaluations'.

Historically, PLAAF's flight education and training has evolved through several modifications since 1949 and is usually divided into five periods – which do not form the scope of this chapter – but help to understand the PLAAF's education and training for cadets at the Air Force Aviation University (AFAU), their follow-up flight training as students at one of the flight colleges, and their transition training at an operational unit.

The current status is closely related to the training system in operation from 2004 to around 2012 and the changes to the flight college organisational structure, which was initiated in 2011 up to the current system of what the PLAAF calls the 'Three Levels and Five Phases' training programme to the 'Four Phase' system. Also, on 1 January 2009 an official 'Outline of Military Training and Evaluation' (OMTE) manual was published, formulating – in fact demanding – fundamental reforms concerning all conditions, particularly under complex EW circumstances and in joint operations in high-technology environments. For this the PLAAF uses the term 'under informatised conditions', which is a concept comparable to the US 'network-centric warfare'. Understanding the general evolution of the PLAAF's aviator and education training system, particularly its progression over the past decade, is important when assessing the system as it stands today.

One major aspect within the current system is related to a reform of the Flight Colleges resulting in a new structure for the consolidated three flight colleges in Harbin, Shijiazhuang, and Xi'an, each of which includes two of the previous flight colleges and has several different types of trainer aircraft rather than a single type of basic and intermediate trainer. This was preceded a few years earlier when the PLAAF reduced the original flight schools – of which there were more than 15 – to only seven flight colleges in the 1990s. In May 2004, the AFAU was founded in Changchun to replace the 7th Flight College that once comprised the Flight Basic Training Base – in fact the former 7th Flight College – and the Flight Training Base.

A true survivor in PLAAF service is the CJ-6A. Remarkably, even though development began in late 1957 and it was first introduced in the mid-1960s, the type is still in production. (Top.81 Forum)

In the end it resulted for the flight academies in a concentration of the different CJ-6 regiments in the former 7th Flying Academy and the JL-8 regiments in the former 3rd Flying Academy, effectively swapping two regiments. In addition, in April 2012, the former 13th Flight College in Bengbu, was transformed into a Flight Instructor Training Base for flight instructors in the three flight colleges and at operational units and was subordinated to the AFAU. The final step not only resulted in the three current flight colleges, but also – and somewhat confusingly for the later ORBAT – it consolidated at least four of the seven Military Region Training Bases (MRTB). The remaining transition training bases were merged with the PLAAF's new air brigades and often re-formed as previously disbanded regiments.

Training syllabus

Given the latest restructurings and including uncertainties or open issues, PLAAF aviator training is, unlike the Naval Aviation training, organised in four phases, which are different for bomber, transport, and helicopter aviator cadets. In 2012, the PLAAF implemented a new 'four-phase' system, which is called the '4+1+1 model' and the '4+1 model', where each number refers to the number of years it takes to complete the programme. For some time new pilots required a total of 10 years to become experienced pilots with the ability to carry out every type of combat mission independently and this long period is one major reason for the revised four-phase structure which cuts this time down to seven years. The four phases are: academic education, professional education, combat aircraft transition training, and combat application training.

Phase 1

It lasts for four years and includes academic education and basic flight training in which the cadets receive their bachelor's degree. The main institution responsible is at first the PLAAF Aviation University in Changchun before the students move to the practical application phase of training at one of the three flight colleges. After the first phase, pilots begin with flight simulation training – which was introduced into the PLAAF training cycle only in the late 1980s but was not in widespread use at operational level – and once a pilot became comfortable with the skills, true flight training started for the last six months in the CJ-6 basic trainer at the university's Flight Basic Training Base. It is not known when the PLAAF decides on which cadets will become pilots or navigators, communicators, aircrew mechanics, or gunners on transports and bombers. It is estimated that this phase totals about 250 flight hours.

Phase 2

It includes one to two years of professional aviation education and transition training at one of the three flight colleges, where the cadets receive transition training onto their final operational aircraft depending on whether they will become pilots for a bomber, fighter, ground attack, helicopter, multi-role type or transport. This decision is made based on an evaluation of pilot performance and ranking of both phases and prepares for Phase 3. Upon completion of the programme, each pilot receives a bachelor's degree in military science and is called a 'double bachelor's' officer. This phase totals from 150 to 200 flight hours.

- Fighter pilots are trained on the JL-8 intermediate trainer for 150 hours (112.3 hours with a flight instructor plus 37.7 solo hours) and 103 hours on the JJ-7 advanced trainer.

- Ground attack pilots fly 150 hours in a JL-8 and 103 hours in a Q-5J trainer.

- Bomber and transport pilots fly 140 hours in a Y-7 trainer but also – since April 2015 – on H-6As (possibly H-6Hs).

- Bomber and transport navigation crew members spend one year receiving theoretical education and training, after which they then train with bomber and transport pilots in a Y-7 trainer.

- Helicopter pilots receive training in a Z-9 helicopter.

Phase 3

It provides flight transition training at an operational unit for the new pilots, which includes technical training, initial flight training and basic tactical training on their future unit's primary combat aircraft. This is similar to the Naval Aviation's syllabus and it seems as if this phase originally lasted between two to three years, but was cut to one, after which the pilots are awarded wings as third-grade pilots. Also, since early

The Sino-Pakistani JL-8 is the PLAAF's standard primary and advanced trainer and was introduced in the late 1990s as a successor to the obsolete JJ-5s. This line-up of JL-8 assigned to the former 3rd Flight Academy, 4th Training Regiment displays the pre-2012 serial number scheme. (CDF)

2011 in this this phase the decision is made if a pilot is selected to become a WSO (in PLAAF parlance a 'rear seat weapons control officer'), which includes five months of theoretical education and training, followed by transitioning into the final unit's aircraft. Most often this is the JH-7A at first but later also the J-16 and Su-30MKK.

Phase 4

It provides (and this differs from the Naval Aviation's syllabus) combat application training at an operational unit in the unit's operational aircraft for another six months. This includes basic tactics training, application of tactics, combined-arms combat and joint combat training. Following completion, the pilots are assigned to their permanent flight squadrons, where they finally move on to their operational aircraft.

PLAAF fighter and multirole combat pilot training

PLAAF			USAF		
Phase	Type	Duration/ Flight hours	Phase	Type	Duration/ Flight hours
Phase 1 basic flight training at Aviation University	CJ-6	6 months/ unknown	Undergraduate Pilot Training (UPT) Phase 1 (Air Force Academy, Officer Training School OTS)	T-6A	6 months/ 79.5
Phase 2 professional aviation education at Flight College	JJ-7, JL-8, JL-9, JL-10, Q-5J	1–2 years/ 150–200	Undergraduate Pilot Training (UPT) Phase 2	T-38	6 months
Phase 3 combat transition training (transition training unit within an operational unit)	JJ-7, JL-9, J-10AS, J-11BS, J-16, JH-7A	1 year/ unknown	Introduction to Fighter Fundamentals (IFF)	T-38, future operational type	3 months
			Formal Training Unit (FTU)	future operational type	7–9 months
Phase 4 combat application training (transition training flight group within an operational unit)	Future operational type	6 months/ unkown			

Evolution and reforms

Besides a general reorganisation and revision of the training syllabuses, there is yet another important issue, which is aimed to dramatically improve the PLAAF's operational readiness through an increase of realistic combat drills and expanding exercises.

For some time one major drawback of the PLAAF's doctrine was its inability to accomplish large-scale exercises within a generally inappropriate and unrealistic operational training. The main reason for this was the lack of an integrated training between different branches like the PLAAF and Naval Aviation and the limited number of aircraft a command post could guide. This did not even change after the overwhelming display of power and its success in the 1991 Gulf War so that certain members within the PLA top brass still did not understand the value of modern aerial combat. This persisted even with the introduction of the Su-27SKs, which were most often operated the same way as J-7s and J-8s, namely, armed with rockets and dumb bombs to destroy ground targets and strictly flown by textbook/manual. Gradually, this thinking of 'just serving support ground units' changed to formulate the aim of 'becoming a force on its own which was capable of running large-scale offensive operations'. This was coupled with the acceptance that the key to any future success lies not only in a dramatic increase of new equipment, but most of all in fundamental changes in training and combat tactics.

A first step in this direction was the establishment of the so-called Flight Test and Training Centre (FTTC) on 1 April 1987, when the former 11th Aviation School at Yanliang was reformed into the FTTC/2nd Regiment and transferred to Cangzhou/Cangxian. This unit is now called Flight Test and Training Base (FTTB) with subordinated brigades. A second important step was the foundation of a specialised test centre for weapons integration, testing and tactical training at Dingxin in June 1999 as a dedicated detachment assigned to the Chinese Flight Test Establishment (CFTE), which has its regular flight test centre in Xi'an-Yanliang.

The extent to which the PLAAF uses modern flight training simulators is little known. Reportedly, they were introduced comparatively late and are rarely documented. This scene taken from a TV report shows a J-10A simulator. (CDF)

Immediately after its establishment at Cangzhou, the FTTC formed three regiments initially flying J-7s and later J-8Bs and Su-30MKKs, dedicated to tactical training for most of the early 1990s – rather than exploring flying techniques or playing an 'aggressor' force – due to its lack of experience in modern tactics. From 1988 onwards, the FTTC received Project Grindstone to simulate a Blue Force aggressor and for most of this period, the regiments were simulating Soviet forces, changing slowly to later simulate ROCAF/USAF. With the political changes, also a cooperation agreement with the Russian base at Lipetsk was signed in order to train the best pilots and controllers at Lipetsk's Red Flag Composite Training and Research Unit. Another base also complementing the FTTC is located at Jiugusheng and flies a J-10 regiment. Concerning the workshare and responsibility, theoretical tactics are usually developed by the PLAAF Command College in Beijing, which are then explored and refined in practice at Cangzhou. In parallel, the development of combat methods begins at an operational unit and ends with testing and approval at Dingxin. In both cases, once these tactical regulations and combat methods are developed and approved, the PLAAF writes the corresponding manuals which all pilots have to study at their operational unit. Commonly, such a process takes at least one year but may last for several years.

Consequently, at the Dingxin Test and Training Base the tactics and flight techniques developed at Cangzhou were explored and verified and one outcome of the

For a long time the JL-8 was regarded as an advanced jet trainer and only in early 2015 did the PLAAF introduce supersonic trainers such as this JL-9 into the advanced flight training phase. This type is probably the final development of the J-7.
(FYJS Forum)

In view of the vast numbers of modern fourth-generation types equipped with digital cockpits it is perhaps surprising that the majority of the PLAAF's trainer fleet – including this JL-8 – is still equipped with analogue instrumentation.
(Top.81 Forum)

early evaluations dramatically showed not only which part of the PLAAF's tactics and training manuals were outdated and needed to be changed, but also the disparity in the training levels and intensities of different forces around the country, clearly showing which units are better trained than others. Since its establishment in 1999, the Dingxin base has been expanded to almost double its original size and now allows training for an entire aviation corps. While being able to handle roughly up to 20 aircraft in its early days, the numbers of aircraft performing the annual exercises and the complexity of simulated war scenarios have increased from year to year. The latest Red Sword exercises now involve up to more than 100 aircraft of different types including fighters and striker and special mission types sometimes up to Theater Command level. This resulted in 2005 in the merger of the three former different test facilities into the current PLAAF test and training base.

Currently, the base features China's first integrated EW training range in Dingxin which is equipped with a fibre-optic network and comprehensive computerised measuring/monitoring equipment including OE and telemetry) to provide real-time information for one of the PLA's digitised command and control centres. Additionally, this base became China's first base to enable realistic training under intensive EM environment, it has tactical air-to-air and air-to-ground ranges, surface-to-air missile (SAM) and anti-aircraft artillery (AAA) positions, radars, simulated enemy command posts (even including a mock-up of a Taiwanese Air Force base), ammunition and fuel depots.

The degree of realistic training in different scenarios has increased so that pilots are now practising flying under challenging environmental conditions, such as at night and in extreme weather conditions. They practice flying at low altitudes over difficult terrain including through valleys, around mountains and over water and also by holding sophisticated multi-branch and inter-service exercises under complex electromagnetic environments and formidable air defence scenarios to mimic actual battle conditions that a potential military adversary may present.

Probably one of the most eagerly awaited types to enter service is the JL-10 advanced jet trainer. It is a true modern trainer that can finally provide realistic flight training for fourth- and fifth-generation types.
(FYJS Forum)

All this has enabled the CFTE and FTTC to run integrated exercises since 2005, the 'Red Sword' and 'Blue Sword' which are comparable to the USAF's Red Flag exercises. While the Red Sword exercises focus on interdiction, CAS, SEAD and OCA (offensive counter air) operations, the Blue Sword exercises are mainly related to aerial combat. Both are aimed at preparing PLAAF pilots for the possibility of future high-technology combat against tactically and technically more advanced adversaries. And, finally, in 2011 the PLAAF established the so-called Golden Helmet air-to-air competition – comparable perhaps to the US Air Force's former William Tell gunnery contest – with the aim of improving and assessing individual pilots' skills and capabilities especially by demanding less scripted free aerial combat manoeuvring. Winning pilots were rated as 'elite pilots' and their units were allowed to proudly wear special markings on their aircraft similar to those taking part in the Russian Aviadarts 2014 competition for the first time. The pilots of the national aerobatics team 'Ba Yi' and 25 additional 'outstanding aviators', were honoured for demonstrating 'unrestricted or free air combat' capabilities and publicity praised to boost their morale.

Another effort in this direction is the PLAAF's aim to actively learn from the flight training programmes of other air forces, especially those in the West. Consequently, in recent years the PLAAF has increased its training with other forces. This includes as a first, the joint Sino-Russian (also including other former CIS states) exercises like Peace Mission (since 2005), and Aviadarts (since 2014), and continuing through several joint Sino-Pakistani exercises called Shaheen (since 2011) up to exercises with Turkey and Thailand. According to several unconfirmed rumours, although the PLAAF performed 'pretty unspectacularly', it rates such exercises as a steady input to learn important lessons in the process; and another positive side-effect is to strengthen its military relationships with foreign states.

Long expected to become the successor of the CJ-6, the CJ-7 trainer, which first flew in late 2010, was eventually put on hold. It might be replaced by a new 'next generation primary trainer' with Hongdu and Guizhou expected to be the main contenders.
(FYJS Forum)

Issues and uncertainties

It seems that the PLAAF has made significant gains by implementing more modern training methods, by revising the structure, streamlining the training process by the increased use of simulators and most of all by employing realistic exercise scenarios. However, as with the Naval Aviation, the biggest weakness is still the lack of adequate training aircraft both in numbers and capabilities. For example, prior to 2011, the JL-8 was regarded as an advanced jet trainer and it was not until early 2015 that the PLAAF introduced supersonic trainers such as the JL-9 into the advanced flight training phase of the flight colleges, which reportedly dramatically reduced the training cycle. Several more JL-9s have been introduced and the PLAAF's aim now is to incorporate the JL-10 trainer into this phase. However, numerically the most important advanced trainer is the JJ-7 but, more seriously, there are at present no adequate primary and advanced turboprop trainers.

Even if this situation can be rectified with the introduction of the modern JL-10, the arrival of this type presents another problem. Due to its much higher flight performance there is – again akin to Naval Aviation – no modern screening trainer in the class of the Diamond DA20 and/or an advanced turboprop trainer such as the Pilatus PC-9 or even the PC-21 to replace the CJ-6 and the JL-8. Furthermore, the current fleet of trainers lacks continuous digital cockpit instrumentation. Before entering an operational cockpit, cadets will only have experienced a digital cockpit in the JL-10.

For some time, the CJ-7 was expected to replace the CJ-6 which, surprisingly, is still in production, but according to recent reports this project has been abandoned and it will be replaced by a more powerful type.

Also, regardless of all efforts to professionalise the training through adherence to a less-scripted but true combat-realistic training with full autonomy over the sorties with little guidance from ground control, which is still several deficiencies remain in the area of combat tactics and skills also in exercises across different branches within

It seems as if the PLAAF has finally decided to re-open the contest for its next-generation primary trainer and both Hongdu and Guizhou are expected to offer submissions. Somewwhat surprisingly, a new private company with close relations to Diamond Aircraft might offer a design based on the Dart 450
(CMA)

The only known image of a
modern Il-76MD/KJ-2000 flight
simulator.
(CDF)

the PLA. In summary, by the time a PLAAF pilot is assigned to an operational unit, about two to three years of practical training as well as roughly four years of academic and theoretical study will have been accomplished, which, compared with the USAF, involves much more time in flight school and associated aeronautics training before arriving at the operational unit. Although reliable numbers are difficult to obtain, an average PLAAF pilot seems to fly less than his or her US counterpart. (PLAAF pilots average about 120 flying hours a year compared to an average of 250 in the USAF). And finally, although the situation has improved dramatically, there are still significant barriers preventing a satisfactory overall development. The main institutional impediment in this regard is the fact that some still deem the PLA to be an army-centric fighting force, so that concepts to evolve a joint command and control system across different branches within the army, navy, and air force are only now gaining currency.

Conclusion

In summary the PLAAF has already undergone – and is still undergoing – an impressive and wide-ranging process of modernisation both in terms of organisation, syllabuses and equipment to train and equip its future combat force under what the PLA itself refers to as 'actual combat conditions' while at the same time reducing the time taken to produce an experienced pilot from 10 to seven years. Most significantly, intensified, regular and realistic war-like exercises have resulted in massive improvements in readiness rates, improved combat effectiveness and also increased confidence and morale within the PLAAF's pilots. The PLAAF is still seeking to cultivate a greater autonomy among its pilots and has begun to shift training away from an emphasis on ground control to a system which encourages independent decision-making. The aim is clear: the current training syllabus is designed to establish it as a technologically advanced, professional, and operationally capable air force to help protect and advance Chinese interests in the Asia-Pacific region and beyond.

PEOPLE'S LIBERATION ARMY AIR FORCE ORDER OF BATTLE 2018

Operational structure

By now after several months of closely following the ongoing PLA reorganisation – officially announced on 18 April 2017 – at least a slightly clearer image becomes slowly visible and consequently it became clear that the PLAAF not only expanded the base/brigade concept initiated in 2011 and adopted it for all Theater Commands but also that this re-organisation is much more profound than the 2012 reform, which laid the fundament of the current reorganisation and which is still far from concluded.

The driving aim behind this reform is most of all to transform the PLAAF into a more capable force, able of joint operations with other Chinese military branches and services. Consequently the PLA initiated its most wide-ranging and ambitious restructuring since 1949 – overall the eleventh since its founding – including an overall major reorganisation of all its services and branches in 2017. Besides the consequences for the PLAAF itself and therefore the main focus of the book, this included the upgrading of the former Second Artillery Force from an independent branch to a new service known as the PLA Rocket Force. Also the PLA Strategic Support Force and the Joint Logistics Support Force were created, as the most significant step made toward enabling the PLA to act as a true joint military force on land, at sea, in the air, and in the space and cyber domains. And this also includes joint training, joint logistics, and joint doctrinal development. Besides the operational drivers, yet other reasons are the desire to improve party supervision over an increasingly complex, corrupt and undisciplined rated system and finally improve the defence R&D system.

Even if not the real focus of this book, before describing the current PLAAF's Aviation Branch organisational structure and reform, it is helpful to take a look at key elements of the new PLA force structure. Still head of the overall PLA structure is the powerful Central Military Commission (CMC), however the former general departments were disbanded and replaced by 15 functional departments, commissions, and offices under the auspices of the CMC. One major reform away from a ground forces dominated army to a joint-operations-capable force is the creation of a separate headquarters for the Army, so that the ground forces now have their own headquarters on par with the other services. The main former seven Military Regions were restructured into five Theater Commands (TC) each one aligned against a certain regional threat in which commanders will be able to command joint forces from army, navy, air force and conventional missile units within their TC. The nuclear part of the former Second Artillery Force (SAF) – which was formally elevated from an independent branch of the army to a true service alongside the army, navy, and air force – still remains at the PLA HQ in Beijing.

The centrepiece of the reform is to create a joint command and control structure with nodes at the CMC and TC levels that will coordinate the PLA's responses to any regional crises and conduct preparations for wartime operations. All this will be accomplished by a reduction of the PLA's overall size by 300,000 men. Most of the reductions will be cut from the ground forces while at the same time an increase in size of the PLAN and PLAAF. So far a major weakness will be overcome by the establishment of a Strategic Support Force (SSF) 'to provide command, control, communications, computers, intelligence, surveillance, and reconnaissance support to commanders and will oversee space, cyber, and electronic warfare activities.' At TC level a new Joint Logistics Support Force (JLSF) will provide logistics support to units.

New structure diagram, Part 1

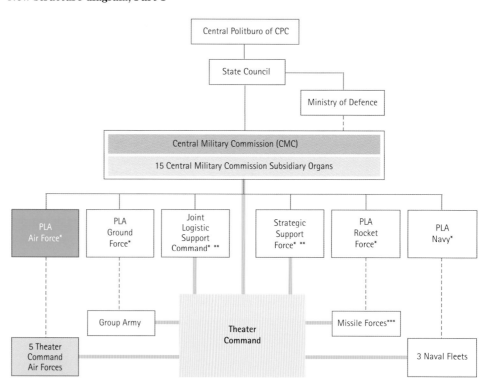

*   All service headquarters execute administrative, but not operational control
**   Support to Theater Commands
*** The nuclear forces still report directly to the CMC.
     For the conventional forces it remains unclear

One relict from the PLA ground forces centric organisation was the structure along Military Regions (MRs) that trained and equipped in general army units and managed provincial-level military districts. The PLAN and the PLAAF were only nominally integrated into this structure and were in practice controlled by service headquarters in Beijing. Consequently, already since the late 1990s Chinese military strategists had discussed reforms to consolidate the MRs into more regionally focused entities, which were briefly called 'war zones' or 'strategic zones', so that at the theatre level, all former seven MRs were replaced by five Theater Commands (TC). A TC is the highest joint headquarters within a respective region, with a primary responsibility for joint training during peacetime but they will exercise command of theatre-based combat forces during wartime. Each of the TC is aligned against a specific 'strategic direction' or respective region and the threats within this sector. Interestingly, only the PLAAF force headquarters are located in the same cities as the TC headquarters, while the Ground Forces and Naval headquarters are located elsewhere.

Within the theaters, army, air force, and naval services – for the East, South and North TCs only – report through two-part chain of command: while administratively, they report to their respective service headquarters in Beijing, they are operationally under the direction of the TC commanders. The individual operational units – for the PLAAF its bases and brigades – are in return under operational and administrative control by their individual service. Despite all modernisations – sometimes resembling a command structure comparable to the US military – the PLA's strict party-rule-reliance is not changing however. With the CMC remaining the military's highest decision-making organ under strict party control, the long-time established 'dual command system' of having a unit commander and a political officer exercising authority, will not be touched.

The PLAAF Aviation Tier

In terms of organisation, the PLAAF is distributed within five branches and – comparable to the Naval Aviation – typically divided between 'aviation', which comprises aircraft and the units operating them, and 'air defence', which consists of ground-based units, such as surface-to-air missiles and anti-aircraft artillery. Chinese publications make a very clear distinction between those two branches.

	PLAAF branches	PLAN branches
1	Aviation	Naval surface vessels
2	Anti-aircraft artillery	Naval submarines
3	Surface-to-air missiles	Naval aviation
4	Radar	Naval coastal defence
5	Airborne Corps (former 15th AC)	Marine Corps

This chapter provides an insight into the organisation and structure of the Aviation Branch, and its elements operating combat and combat support aircraft. Even after these wide-ranging reductions, with a strength of still over 2,500 aircraft deployed across the PRC and responsible not only for air defence but also for airborne early

warning, reconnaissance, strategic and anti-ship operations, tactical air support, transport and other support operations, the Aviation Branch of the PLAAF is the most important body within Chinese military aviation.

Long structured along PLA lines, the Aviation Branch is organised into Operational Areas, also known as Air Force Districts (AFDs). Since 1985, each of the AFDs has mirrored the so-called Military Regions (MRs) of the PLA, which are generally based according to the political and administrative organisation of the various provinces of the PRC, and therefore vary in size depending on the geographic area and its strategic importance. Each of seven AFDs exercises direct control over its respective Military Region Air Force (MRAF). The MRAFs of the PLAAF were under the dual control of the Air Force and the relevant Military Region of the PLA. This changed with the revised structure of Theater Commands and consequently the PLAAF is currently organised into five AFDs, which therefore operate five Theater Command Air Forces (TCAFs). The manner in which these are usually presented in official Chinese publications is known as the Official Protocol Order and this order plays an important role in Chinese military hierarchy and essentially dictates the way in which the entire PLAAF structure is presented in this chapter as well.

- Eastern TC (ETC) takes responsibility for the Taiwan Strait and East China Sea.
- Southern TC (STC) is focused on the South China Sea.
- Western TC (WTC) is responsible for the Sino-Indian border and Central Asia especially cross-border terrorism.
- Northern TC (NTC) handles challenges emanating from the Korean Peninsula and Japan.
- Central TC (CTC) is responsible for the defence of the capital and will provide support to other TCs in case of urgency.

A recent change however is, that each TCAF no longer entirely exercises control over its subordinated units via Command Posts (CPs). Historically, the PLAAF has had a five-tiered command structure – if including down to the lowest operational level even seven tiers (PLAAF HQ – MRAF HQ – command post – air division – air regiment – flight group – flight squadron) for its aviation troops, but in order to simplify and streamline the chain of command was reorganised in the course of PLAAF reorganisations during 2003-4, when the number of tiers was reduced to four – or similar a six-tiered structure (PLAAF HQ – MRAF HQ – base – air brigade – flight group – flight squadron). The first change was the abolishment of the air corps level for command headquarters and the creation of several command posts (CPs) as described in *Modern Chinese Warplanes* (Harpia Publishing, 2012). The operational chain of command was consolidated under the seven former MRAF Headquarters. When the next round of changes was initiated in early 2011, four bases were created from existing CPs.

For the vast majority of combat air units – either brigades or divisions – the structure follows the system below, and consequently a CP might substitute for a base and an air division might substitute for a Brigade. Even if not entirely clear, it seems as if in early 2018, only three CPs remain active – Zhangzhou in the ETC, Hetian in the WTC and Changchun in the NTC – but it appears that no brigades are subordinate to them but solely to their bases.

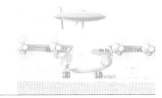

New structure diagram, Part 2

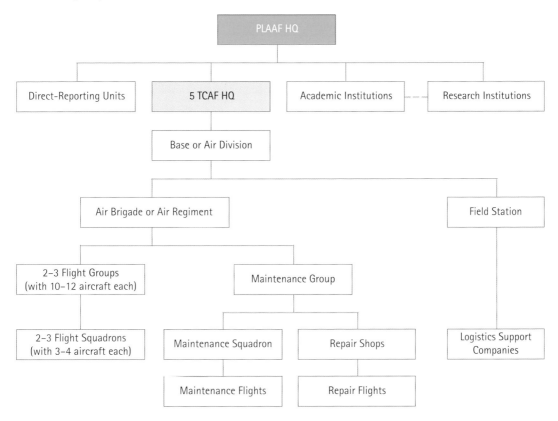

2011 and 2017 Reform: bases and brigades

In standard PLAAF terminology, the Aviation Branch is usually described as consisting of the following types of units:

- Fighter (jianjiji hangkongbing)
- Ground Attack (qiangjiji hangkongbing)
- Bombers (hongzhaji hangkongbing)
- Transports (yunshuji hangkongbing)
- Reconnaissance (zhenchaji hangkongbing)

For many years – until 2012 – the best-known hierarchical units within the PLAAF tier have been the air divisions and their corresponding air regiments, which formed the most suitable starting point for an overall review. Additionally each CP controlled a number of subordinated ground-based assets (including SAM brigades and regiments, AAA regiments, radar, communications and other support units), as well as several of the most important combat formations of the PLAAF: the former air divisions (ADs) and air regiments (ARs), which were mostly disbanded and superseded by the concept of bases and air brigades (ABs).

In late 2011 the PLAAF initiated its last major unit reorganisation which included the introduction of bases and brigades in January 2012, when four bases (as corps deputy leader-grade) were formed from existing CPs within only four of the seven MRAFs – Guangzhou, Lanzhou, Nanjing and Shenyang. These four – Dalian, Nanning, Shanghai and Urumqi – execute command and control of their subordinated air brigades but also SAM, AAA, and radar units in their area of responsibility (AOR) and they are also responsible to coordinate joint training with Army and Navy assets in their AOR. Quite noteworthy, this shift to an air brigade structure occurred parallel not only in operational units but also in the PLAAF's flight colleges, when the seven flight colleges were merged into three flight colleges at Harbin, Shijiazhuang, and Xi'an and one Flight Instructor Training Base subordinated to the Air Force Aviation University.

The reform including the creation of brigades and bases was put on hold at the end of 2012 and even if it is not entirely clear why there was a delay, the reform was re-instituted in early 2017. In the meantime – since early 2016 – the former seven MRAF HQs were reorganised into five TCAF HQs. Even now, the reason for the near-five-year hiatus in creating new bases and air brigades is not entirely understood. However, it was likely to gain experience with the new structure and its operational benefits. Thereafter, the PLAAF restarted this rapidly progressing reform with a much larger reorganisation in April 2017. For several months, a new brigade was confirmed each week, aircraft were renumbered, reports about newly established bases appeared and older divisions were either restructured or disbanded. Meanwhile, their former regiments were either converted to brigades or disbanded – the latter above all in the case of those equipped with obsolescent aircraft. By now there are usually two bases within each TC, which command the individual brigades in each of the five formal TCAFs.

Summary – PLAAF Bases (up to September 2018)

Current Theater Command	Former Military Region (MR)	Former Command Posts			Confirmed Bases
		MR Deputy Leader	Corps Deputy Leader	Division Leader	
Eastern	Nanjing		Fuzhou	Shanghai Zhangzhou	Fuzhou, Shanghai
Southern	Guangzhou			Nanning	Kunming, Nanning
	Chengdu	Lhasa	Kunming		
Western	Lanzhou	Lanzhou	Urumqi (Wulumuqi) Xi'an	Hetian	Lhasa, Lanzhou, Urumqi
Northern	Jinan	Jinan			Dalian, Jinan, unconf. location
	Shenyang		Dalian	Changchun	
Central	Beijing		Datong		Datong, Wuhan
	Guangzhou		Wuhan		

As of mid-July 2018 – and after several months of continuous changes – the speed of newly established brigades and their superordinated bases slowed down. Even if the reform is clearly still ongoing, it seems as if the final structure has at least become

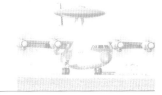

visible. Consequently it seems as if the PLAAF not only expanded the base/brigade concept and adopted it for all Theater Commands but also that this re-organisation is much more profound than the 2012 reform. So far it seems as if the PLAAF entirely converted to this new organisation across its tactical combat units. For the strategic assets – the bomber, transport and specialised EW and reconnaissance units – however it seems as if these divisions and regiments will survive and also the Naval Aviation has transformed its former naval air divisions in a similar way to naval air brigades.

Besides the mentioned tactical operational combat brigades and the training brigades, the PLAAF created several more brigades: these are so called Transport & SAR Brigade in each TC and one specifically assigned to the PLAAF's HQs, additionally several unmanned aerial vehicle (UAV) brigades were formed and the former FTTC, TTC and CFTE trails and evaluation regiments were converted to brigades. Also the PLAAF's Airborne Force formed a dedicated transport aircraft aviation brigade, while at the same time changing its entire force to a brigade structure similar to the reorganised Group Armies of the PLA Army/Ground Forces. Anyway there are still many uncertainties especially concerning which brigade is subordinated to what base and consequently this ORBAT is only an attempt to represent the current situation.

Before describing the PLAAF ORBAT it is necessary at first to differ between an air base and a base. An air base is simply an operations centre for units of an air force, on which a certain unit is based. However, a base is a leader-grade organisation. So even if at first sight it looks as if regiments were replaced by brigades usually just by reusing their original numbers and the divisions are replaced by bases, the difference is most of all related to its leader grade: each base – one could probably better say a command base – is directly subordinate to the relevant Theater Command Air Force HQ, which is a Theater Command deputy leader grade organisation. As such, each of the bases has command overall all PLAAF air brigades (division deputy leader grade), AAA, SAMs, and radar units in their immediate AOR. Usually each of these bases also have associated, specific forward operating bases assigned and perhaps a base even commands the supporting infrastructure like airfield units, construction units and mobile units that can prepare unused airfields to turn them into operating airfields.

For long and depending on the type of operational unit, they were sub-divided into subunits with divisions (*shi*) and regiments (*tuan*) and regiment-grade field stations, battalion-grade flight and maintenance groups, and company-grade flight and maintenance squadrons. Operationally the regiments are subdivided into groups (*dadui*) and squadrons (*zhongdui*) as the most important combat assets. Concerning the number of aircraft assigned to a brigade, the number will most likely not change in the immediate future even if in the longer term, it might be increased. The reason for this is that usually a typical fighter/attack air regiment has 18-24 aircraft – bomber and special mission units are usually smaller by numbers – while an air brigade can have 18-40. These 18-24 aircraft are usually organised in three flight groups with six to eight aircraft each for a regiment, whereas a brigade can have even four or five flight groups (*daduis*). Complementing these regular operational types, a brigade usually is also equipped with one or two flying squadrons (*zhongdui*) of trainers. There are a few report that assume as if originally it was even planned to established multi-functional or multi-role brigades in which one brigade is equipped with each flight group equipped with a different type of aircraft, namely operational fighters and/or ground-attack or multirole and trainers. However, so far only the 172nd and 176th Air Brigades have such a complement and it appears as if this approach has been abandoned after the five to six years of gaining experiences in favour of the current structure. Reasons for this might be different and especially maintenance issue due to the higher complexity of operating small numbers of different types in one unit is. So in consequence by 2018 all brigades have only the three flight groups inherited from the former air regiment they upgraded from. Another reason is, that some types – like the JH-7A, J-10A and J-11A – are no longer in production or in case of the Su-30MKK only a limited number were imported, so that existing brigades cannot gain additional aircraft without abolishing at least one to spread its aircraft over the remaining brigades.

Chinese designation	No. of aircraft	Chinese	Pinyin	USAF equivalent
Air Division	3x 24	航空兵师	Hangkongbing Shi	Air Wing
Air Brigade	24–28	航空兵旅	Hangkongbing Lv (Lü)	Squadron
Air Regiment	24	航空兵团	Hangkongbing Tuan	Squadron
Independent Air Regiment	n/k	航空兵独立团	Hangkongbing Duli Tuan	
Flying Group	8	空大队飞行大队对	Feixing Dadui	
Reconnaissance Group	n/k	侦察大队	Zhencha Dadui	
Flying Squadron or Subunit	4	飞行中队空中队	Feixing Zhongdui	
Flight Unit/Detachment	12–20	部分队	Fendui	

So far, no twin-seat variant of the updated second-generation J-10 exists. J-10B/C brigades instead employ a few J-10AS trainers each. This pair is assigned to the 5th Brigade. (Fan Yishu via chinamil.com)

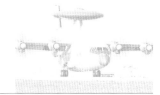

Issues & limitations

It is interesting to follow these latest rounds of reorganisations, especially since earlier reports that these efforts would pause or even end at the end of 2017 proved incorrect. Also, it remains to see what will happen with the remaining divisions. It is possible they will also convert to the base/brigade system at a later stage or they may remain as they are. Even more intriguing are the reasons why regular combat units for fighters and multirole types appear better suited to the new system and why bombers, transports and special mission units have remained assigned to divisions. Otherwise – and most similar to the Naval Aviation – one of the most often mentioned concern or still system-immanent limitations is that several issues between the PLAAF and the Naval Aviation remain unclear. A question remains as to the true 'jointness' of joint operations and consequently there are very differing opinions about the overall effectiveness of the PLAAF as a joint fighting force.

The biggest concern – also within circles of the PLAAF leadership – is therefore on how willing the PLA leadership is to change. Still today the majority of the PLA remains as ground forces-centred army, the majority of the leaders throughout the newly-created CMC 15 subordinate organisations are ground force officers and only one of the five TC commanders is a PLA Navy officer and again only one is a PLAAF officer even if their command staffs have truly become joint with a mix of Army, Navy, and Air Force officers. Overall – and as mentioned in the introduction of the chapter as one driving reason for the current reform – the PLA itself is seen as an increasingly complex, corrupt and undisciplined system that remains most conservative and excessively concerned with officer grade and 'protocol order' in decision making. In the end, the question remains at which pace the PLAAF will becomes an effective part of a truly joint PLA?

This image offers very rare confirmation that the PLAAF also acquired several of the longer-range R-27TE2 missiles. Unfortunately the serial number of the Su-30MKK is removed to hide its assignment.
(FYJS Forum)

ORBAT 2018

The following pages attempt to reconstruct the order of battle of the PLAAF as of September 2018, so exactly six years after the original *Modern Chinese Warplanes*.

After the latest PLA military reform, the PLAAF is now divided into Theater Commands and their respective Theater Command Air Forces.

Key

● Theatre Command HQ

● Base

○ Former Command Post

A map of China showing the five Theater Commands. (Map by James Lawrence)

The five new Theater Commands in protocol order:

- Eastern Theater Command (东部战区)
- Southern Theater Command (南部战区)
- Western Theater Command (西部战区)
- Northern Theater Command (北部战区)
- Central Theater Command (中部战区)

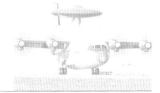

Eastern Theater Command

Successor to the former Nanjing Military Region Air Force (Nanjing jungu kongjun/ Nankong), the current Eastern Theater Command and its ETCAF were founded on 1 February 2016. The former Nanjing MRAF came into being through the reorganisation of the Huadong Military Region Air Division in August 1950. Only a month later, this MRAF redeployed to Shanghai and merged its forces with the Shanghai Air Defence HQ, changing its designation to the Huadong MRAF (eastern China) in the process. In November 1954, this MRAF was again re-designated as the Huadong Military Region Air Force Department, and in June 1955 it became the Nanjing MRAF. In September 1957, the Nanjing Military Region Air Defence Forces were incorporated within its structure, but in July the following year a considerable portion of its assigned units – primarily the 1st and 5th Air Corps – were separated to form the Fuzhou MRAF, in Jinjiangxian. The Fuzhou MRAF was disestablished in September 1985, and its forces merged back into the Nanjing MRAF.

The primary importance of the ETC lies in its proximity to Taiwan and the East China Sea facing Japan. As a result, this area includes a significant number of major bases for PLA ground forces and missile units, as well as hosting the main base of the East Sea Fleet of the PLAN. Its jurisdiction includes the ETC Naval Fleet – the former East Sea Fleet – and three reorganised Group Armies (GA): the 71st GA (former 12th GA), the 72nd GA (former 1st GA) and the 73rd GA (former 31st GA). Its joint command headquarters are located in Nanjing.

Its Eastern Theater Command Air Force (ETCAF) is responsible for air defence of the provinces of Anhui, Fujian, Jiangsu, Jiangxi, Zhejiang and the city of Shanghai.

Major units currently assigned to the Eastern Theater Command Air Force (ETC AF, 东部战区空军) are described on the following pages.

The most potent long-range aircraft in PLAAF service is the latest H-6K, like this example flown by the 28th AR. This particular aircraft took part in the Aviadarts 2018 exercise in Russia.
(Giovanni Colla)

10th Bomber Division

The 10th Bomber Division is one of the oldest units within the PLAAF and was established on 17 January 1951 on the basis of elements from the former 66th Division PLA, at Nanjing. Initially equipped with Tupolev Tu-2 bombers, this division participated in the Korean War, before being re-equipped with Tupolev Tu-4 bombers and redeploying to Nanjing, in the 1960s. It gained indigenous H-6A bombers later and participated in several nuclear bomb tests during the 1970s and 1980s. In 2012 the 30th Air Regimnent (AR) had been disbanded, its base Nanjing/Dajiaochang closed and the Y-8s transferred to re-establish the 20th Division. In parallel the 28th AR gained the first H-6K and transferred the older H-6H to the replace H-6A within the 29th AR in 2013. By 2017 the 30th AR was re-established again gaining H-6M from the 108th AR so that currently this division operates all the H-6 air force variants.

A rare image showing both 'legacy' H-6 variants at Luhe/Ma'an: the standard H-6H armed with KD-63s in the foreground is operated by the 29th AR; the H-6M is from the re-established 30th AR, which gained its bombers in 2017, when the 108th AR received the H-6K.
(CDF)

26th Specialised Division

Initially established as the 26th Fighter Division at Liuzhou, in December 1952, this division was equipped with Soviet-made Lavochkin La-9 and La-11 fighters. It later gained J-6 interceptors and was redeployed to Yunan Province in April 1966, to boost PRC air defences during the Vietnam War. In 2005-06, the 26th Division was reorganised as a Special Mission unit equipped with airborne early warning and other Y-8 special mission aircraft. In 2017 the division was reorganised with now only two regiments left, when the 77th AR dedicated to search and rescue duties was turned into a dedicated transportation and search and rescue brigade.

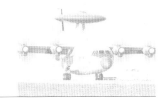

In 2017 the 76th AR assigned to the 26th Special Mission Division was split into two, reorganised together with the former 77th AR and all aircraft renumbered. The 76th AR flies the KJ-500 – the most numerous PLAAF AEW asset. (DS via SDF)

The Y-8Ts (GX-4s) were most likely all reassigned to the 77th AR and operate from the unit's main air base at Wuxi-Shuofang together with the KJ-200, updated KJ-200A and KJ-2000. (DS via SDF)

In December 2016 the first KJ-200 was upgraded to KJ-200A standard featuring a new nose-mounted AEW radar. This variant now flies within the 77th AR and is seen during the Red Sword 2018 exercise. (Yang Jun via chinamil.com)

Fuzhou Base

In line with the second round of establishing bases and brigades, the Fuzhou base was formed as the second base within the ETC. Today, this base more or less is the successor to the former 14th Fighter Division additionally is responsible for the UCAV brigade.

Fuzhou itself as an air base was long home of the former 49th Division flying J-6 and still today it houses one of the several UCAV daduis.

40th, 41st and 42nd? Brigades

Currently the status of the three former regiments is only confirmed for the 40th and 41st Brigades. The 14th Fighter Division was established in February 1951 at Beijing Nanyuan and after operating J-5s and J-6s during the 1950s and 1960s, the unit was re-equipped with J-7s in the 1980s.

During the first round of reorganisations, the original 41st AR was transferred to become the 86th Brigade in 2012. However, it was soon replaced by the former 86th AR, 29th Division, which was renumbered as the 41st Air Regiment in due course then moving to Wuyishan. Subsequently, both the 40th AR and 41st AR were re-equipped with J-11As and Su-27UBKs and by around June 2017 both were turned into brigades. Since December 2017 the 40th Brigade is receiving J-16, with its J-11s going to 55th Brigade in the CTC. The status of the 42nd AR is still unclear.

85th Brigade

Similar to its sister brigade, the 86th, the 85th was one of the first brigades established in 2011 and originally assigned to the Shanghai Base in 2012. Again, it too was once assigned to the 29th Fighter Division as the former 85th AR and originally flew J-7B until 2003, when they were replaced with Su-30MKK.

Quite surprisingly in 2017 in line with the second round of changes, assignment changed to the newly established Fuzhou Base.

180th Brigade – Unmanned Attack Brigade

This unit was established by around 2011-12 and represents a combat formation that is probably unique in the world. It consists of several squadrons operating old Shenyang J-6 fighters that have converted into unmanned combat aerial vehicles. Its organisational structure is still unclear since overall it consists of several subunits (fendui) dispersed over several airfields.

Not much is known about the unique J-6W UCAV variant of the J-6 fighter: these two are assigned to the 180th Brigade, 71st Fendui at Jinjiang. (CMA)

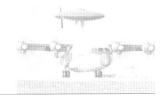

The 40th Brigade was long a J-11A operator but in December 2017 received its first J-16s in low-visibility markings; the J-11s transferred to the 55th Brigade.
(Wang Yijie via chinamil.com)

Both former 14th Fighter Division regiments flew J-11 fighters for many years. While the 40th Brigade has upgraded to the latest J-16, the 41st Brigade is still flying J-11As.
(Top.81 Forum)

Su-30MKKs were delivered to the then 29th Division, 85th AR, in 2003. This became a brigade in 2012 but changed subordination from Shanghai Base to Fuzhou Base in 2017.
(DS via CDF)

Shanghai Base

Within the brigade reorganisation introduced in 2012, the former 29th Fighter Division was brought under the command of the Shanghai Base, with its HQ at Quzhou, and associated regiments were expanded to brigades.

In 2017 the Shanghai Base was expanded again and reorganised and gained several more brigades stemming from the former 3rd, 14th, 28th, 29th and 32nd Divisions.

7th, 8th and 9th Brigades

The final round of expansion occurred in 2017, when the 3rd Division was disbanded and its regiments transferred as brigades to the Shanghai Base. This division – together with the 1st and the 2nd Fighter Divisions – was one of the oldest and most decorated units of the PLAAF. This might be a reason that quite unusually all three regiments were transformed into brigades in June 2017. The 3rd Division was originally established with Soviet assistance from the former 3rd Pursuit Brigade on 7 November 1950 and was involved in the Korean War from October 1951, where it distinguished itself in the course of several air combats. During the 1950s, the 3rd Fighter Division took part in clashes with Nationalist forces along the eastern coast of the PRC, and in the 1960s was redeployed to the south and southeast coasts of China, where it saw engagements with USAF fighters and UAVs over Hainan Island. Of these three brigades, the former 7th AR flew J-7B until 2011, which were replaced by J-7E/L but already in 2017 this unit gained J-16, with its J-7L going to 21st Brigade.

The 8th Brigade replaced its J-7B in early 2006 with J-10A/AS, which are still operational and finally the 9th Brigade replaced its original Su-27SK/UBK from the 1990s with Su-30s at around 2001. The 9th Brigade is said to become the first PLAAF frontline brigade to operate the J-20A.

Another Su-30MKK unit active within the ETC since 2002-03 is the former 3rd Division, 9th AR, which replaced the Su-27SK/UBK. It became a brigade in 2017 and one of its aircraft is seen here during the Sino-Russian Golden Dart 2018 exercise firing a B-13L rocket pod, still the most widely used air-to-ground weapon.
(Yang Pan via chinamil.com)

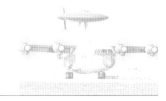

A former J-7L unit, the 7th Brigade received J-16s in 2017 but they were only confirmed by images in mid-2018.
(CCTV-7 via CDF)

The J-10As operated by the 8th Brigade were delivered in early 2006 replacing J-7IIs within the former 8th AR, which became a brigade in June 2017.
(KJ.81.cn)

The 9th Brigade's Su-30MKKs are the oldest examples of the type in service. In May 2018 rumours emerged that the 9th Brigade might become the first frontline unit to receive the J-20A.
(Top.81 Forum)

78th Brigade

In 2012 the 78th Brigade was formed flying J-8DHs, before these were replaced in 2017 by modified J-8DFs (left). Due to the lack of J-8 trainers, these units usually operate a few JJ-7As (right).
(both Top.81 Forum)

The 78th Brigade too has a mixed and quite confusing history, since it originally stems from the 29th FD, 87th AR that operated out of Chongming flying the J-8B. This until was later transformed into special operations regiment but reassigned in 2012 to the Shanghai Base as the 78th Brigade. In 2017 the old J-8DHs were replaced with modified J-8DFs coming from the former 4th AR.

83rd and 84th? Brigades

The ETC operated two dedicated strike units for some time: the former 83rd and 84th ARs once assigned to the 28th Ground Attack Division. The 83rd AR reappeared as a brigade in mid-2017, but the fate of the 84th AR is still unconfirmed.
(Left: DS via SDF,
right: hunter_chen via Top.81 Forum)

Also a new addition gained in 2017 is the 83rd Brigade and eventually the 84th too, both from the former 28th Ground Attack Division. Very little is known about this division, except that it was established at Gucheng in December 1952 and is often regarded as the most prestigious PLAAF attack unit. All of its regiments flew Q-5s until 2003-4, when the 83rd AR exchanged its Q-5Bs for JH-7As, while the 84th AR replaced its Q-5Ds with JH-7As in late 2011. The 82nd AR has since been re-equipped with LGB-capable Q-5Ls, but quite surprisingly this most capable Q-5 variant was retired first and its regiments – like the 82nd AR – disbanded in March/April 2017. The 83th AR became a brigade in June 2017 but the current status of the 84th AR is unconfirmed.

86th Brigade

The 86th is in fact one of the first brigades established in 2011 and even if as a predecessor forming the basis of the Shanghai Base, the former 86th AR was assigned to the 29th Fighter Division it has quite a mixed history. The 29th FD was established in January 1954 at Jiaxing and spent most of its history equipped with J-6s facing Taiwan. The strength of its regiments was boosted significantly through the introduction of

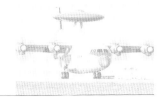

A fine study of a J-7E assigned to the 86th Brigade, which was one of the first brigades confirmed in April 2012 (Top.81 Forum)

Su-30MKKs to the 85th Brigade – however transferred to the newly formed Fuzhou Base in 2017 – and J-7E gained in April 2012, when the 86th Air Brigade was formed at Jiujiang/Lushan from the former 14th AD/41st AR and moved to Rugao in early 2013.

93rd Brigade
Similar to the 78th Brigade, the 93rd Brigade too has a mixed and quite confusing history. It started life in 2005 as the 3rd Independent Regiment, which was subordinated to the 26th Division as the newly formed 78th AR. It gained J-8FR replacing JZ-6 at around 2011 and in April 2012 the 78th AR was transformed into a brigade subordinate to Shanghai Base.

Besides two dedicated reconnaissance regiments, the PLAAF's 93rd Brigade has flown the JZ-8F since 2011. (Top.81 Forum)

95th Brigade

The 95th Brigade was formerly the 32nd Division, 95th AR, a unit that was only re-established in early 2012 through the combination of units assigned to the former MR Training Base with regiments assigned from other divisions. In 2017 the 32nd Division was disbanded and quite surprisingly the newly formed 95th Brigade transferred to the Shanghai Base, replacing the already established 85th Brigade, which was moved to the newly established Fuzhou Base.

The J-11Bs assigned to the 95th Brigade are rarely photographed and images showing their serial numbers are even less common. (Top.81 Forum)

UAV Battalion?

The ETC also operates one dedicated UAV reconnaissance battalion with two regiments operating from two dispersed airfields. The unit at Daishan was re-established in 2013 when this disused air base was reopened for use by UAVs for airborne early warning duties; the other base is located at Ningbo.

Since around 2013 the PLAAF has also operated the BZK-005 for AEW duties off Daishan Island, most likely together with Naval Aviation. (CDF)

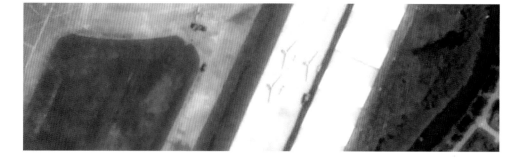

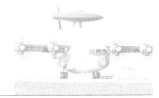

Eastern Theater Command HQ Flight and SAR Brigade

In line with the 2017 reorganisation, the PLAAF established dedicated Theater Command transport and search and rescue (SAR) brigades operating a mix of Z-8K, Z-9B and Mi-171V-5/7 helicopters and/or Y-5C and Y-7 fixed-wing transports.

The HQ Flight and SAR Brigade reports directly to the Eastern TCAF Headquarters.

The two best known types flown by the ETC Transport and Search and Rescue Brigade are the Mi-171 (left) and Z-8KA (right) helicopters which were formerly flown by the 77th Regiment's SAR detachment.
(both DS via CDF)

Aviation units assigned to the Eastern Theater Command

Code	Unit (Division/Regiment)	Base	Aircraft type	Remarks
	10th Bomber Division			HQ Anqing
20x1x (01-49)	28th Air Regiment	Anqing	H-6K	Air Base also known as Anqing North
20x1x (51-99)	29th Air Regiment	Luhe/Ma'an	H-6H	
21x1x (01-49)	30th Air Regiment	Luhe/Ma'an	H-6M	Formerly based at Dajiaochang?; gained H-6M from 108th AR in 2017
	26th Specialised Division			HQ Wuxi-Shuofang
3007x 30x7x 3027x	76th Airborne Command and Control Regiment	Wuxi-Shuofang	KJ-500 (GX-10) Y-8C Y-8T (GX-4)	Aircraft were renumbered in mid-2017; Y-8C and Y-8T status unclear
3057x 3067x 3087x	77th Airborne Command and Control Regiment	Wuxi-Shuofang	KJ-2000 KJ-200A (GX-5) Y-8T (GX-4)	Aircraft were renumbered in mid-2017
30x7x (51-99)	77th Air Regiment (Det.)	Nanjing-Daiiaochang	Y-7-100, Y-7G	Air base also known as Daxiao; status unclear, most likely merged into TC SAR brigade
???	???	Jiujiang/Lushan	Y-9?	Since 2017, a new refurbished base for transports, also known as Mahuiling
	Fuzhou Base			HQ Fuzhou
65x1x	40th Air Brigade	Nanchang-Xiangtang	J-11A, J-16, Su-27UBK	Former 14th AD/40th AR, 20x5x (01-49); under conversion to J-16
65x2x	41st Air Brigade	Wuyishan	J-11A, J-11BS	Former 14th AD/41st AR, 20x5x (51-99)
65x3x	42nd Air Regiment	Zhangshu	J-7L	Former 14th AD/42nd AR, 21x5x (01-49); status unclear; reportedly lost in 2017 J-7L to gain J-11B/BS
69x6x	85th Air Brigade	Quzhou	Su-30MKK	Reportedly reassigned from Shanghai to Fuzhou Base
79x1x	180th Unmanned Attack Brigade			
?	1st Dadui	Liangcheng Longyan Guanzhi	J-6W/B-6	A forward operational base (FOB)
?	2nd Dadui	Yangtang Li	J-6W/B-6	Located within the STC
?	3rd Dadui	Wuyishan	J-6W/B-6	
?	4th Dadui	Ji-an/Taihe Liancheng	J-6W/B-6	Air base also known as Jinggangshan
?	5th Dadui	Fuzhou	J-6W/B-6	

Note: Organisational structure of this unit is unclear since overall it consists of several subunits (fendui) dispersed over several airfields (known so far are):

60F = Xingning
61F = Liangcheng
70F = Wuyishan
71F = Jinjiang (Quanzhou-Jinjiang)
75F = Hui'an (Luocheng/Huian)
80F = Longtian, also FOB
85F = Fuzhou

	Shanghai Base			HQ Quzhou
61x8x	7th Air Brigade	Wuhu	J-16	Former 3rd AD/7th AR, 10x4x (01-49)
61x9x	8th Air Brigade	Changxing	J-10A/AS	Former 3rd AD/8th AR, 10x4x (51-99)
62x0x	9th Air Brigade	Wuhu	Su-30MKK	Former 3rd AD/9th AR, 11x4x (01-49); reportedly first unit to gain J-20A
68x9x	78th Air Brigade	Shanghai-Chongming Island	J-8DF, JJ-7A	
69x4x	83rd Air Brigade	Hangzhou-Jianqiao	JH-7A	Former 28rd AD/83rd AR, 30x9x (51-99)
31x9x (01-49)	84th Air Regiment (?)	Jiaxing	JH-7A	Former 28rd AD/84th AR; status unconfirmed, reportedly also a brigade
69x7x	86th Air Brigade	Rugao	J-7E, JJ-7A	
70x4x	93rd Air Brigade	Suzhou	JZ-8F, JJ-7A	Air base also known as Suzhou-Guangfu
70x6x	95th Air Brigade	Lianyungang, Baitabu	J-11B/BS	Former 32nd AD/95th AR, 40x3x (51-99);
	UAV Battalion			
?	1st UAV Regiment	Ningbo/Zhuangqiao	BZK-005, BZK-006?	A few JH-7As were seen there as well
?	2nd UAV Regiment	Daishan	BZK-005, BZK-007	
	ETC HQ Flight			**Hong Kong/Shek Kong**
51x1x	Eastern Theater Command Transport & SAR Brigade	Nanjing City/Lukou	Mi-171V-5, Y-5C, Y-7, Z-8K, Z-9B	Most likely former 77th AR SAR Detachment

Forward operational bases at: Jinjiang, Longtian, Luocheng/Huian, Macuoping, Quanzhou (Jinjiang Intl.), Xiapu and Zhangzhou

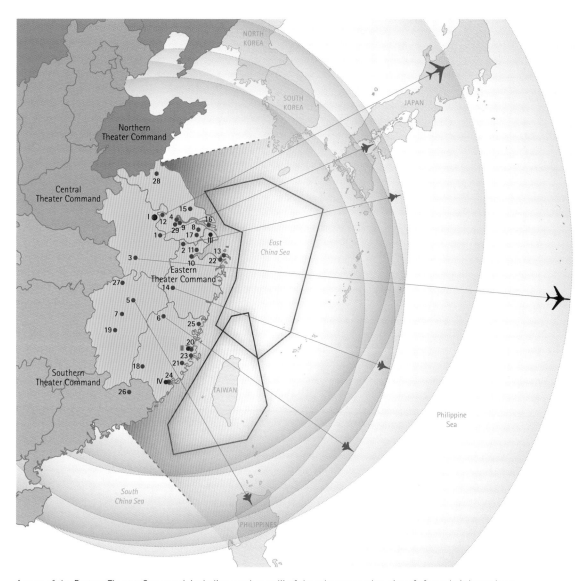

A map of the Eastern Theater Command, including combat radii of the relevant combat aircraft from their home bases.
(Map by James Lawrence)

Key

● Theatre Command HQ

● PLAAF Command Post

○ Former Command Post

● PLAAF Air Base

— China ADIZ

— Taiwan ADIZ

PLAAF Air Bases

I Nanjing HQ
II Fuzhou
III Shanghai
IV Zhangzhou
1 Wuhu (HQ)
2 Changxing
3 Anqing North (HQ)
4 Nanijing-Dajiaochang
5 Nanchang Xiangtang
6 Wuyishan
7 Zhangshu
8 Wuxi-Shuofang
9 Nanjing-Tushan
10 Hangzhou-Jianqiao
11 Jiaxing
12 Luhe/Ma'an

13 Daishan
14 Quzho
15 Rugao
16 Shanghai-Chongmin
17 Suzhou
18 Liangcheng Longyan Guanzhi
19 Ji-an/Taihe Liancheng
20 Fuzhou
21 Luocheng/Huian
22 Ningbo/Zhuangqiao
23 Longtian
24 Zhangzhou
25 Xiapu
26 Yangtang Li
27 Jiujiang/Lushan
28 Lianyungang/Baitabu
29 Nanijing City/Lukou

Combat radii

JH-7A	900km (586nm)
J-10A/AS	1,300km (702nm) with tanks
J-11A/B, J-16, Su-30MKK	1,340km (724nm)
H-6H/M	1,800km (972nm)
H-6K	2,500km (1,350nm)

Southern Theater Command

Successor to the former Guangzhou Military Region Air Force (Guangzhou jungu kongjun/Guangkong, the current Southern Theater Command and its STCAF were founded on 1 February 2016. The Guangzhou MRAF was established in July 1955 from units previously under the control of the Zhongnan Military Air Force Department (south-central China), itself created in September 1950, originally as the Zhongnan Military Region Air Division. In May 1957 the Guangzhou MRAF assumed control of the Guangzhou Military Region Air Defence Force, followed in September 1985 by all units of the disestablished Wuhan MRAF. Ever since, the Guangzhou MRAF has been responsible for air defence of the provinces of Guangdong, Hainan, Hubei and Hunan, as well as the autonomous region of Guangxi, the Hong Kong Special Administrative Region and Macau.

In contrast to the ETC, which remained as its predecessor, the former Guangzhou MR lost responsibility for the Hubei province, which was assigned to the Central Theater Command, but it gained responsibility for the Guizhou and Yunnan provinces from the former Chengdu MR. The STC's jurisdiction includes the two reorganised Group Armies (GA): the 74th GA (former 42nd GA) and the 75th GA (former 41st GA) and its joint command headquarters are located at Guangzhou. Consequently the STC is responsible for the STC naval fleet (the former South Sea Fleet). Since 2003-4 it has exercised control over subordinated units via CPs in Nanning and Wuhan. The STCAF is the second of two PLAAF commands facing Taiwan, and strategic importance is added by the fact that it also faces the Philippines and Vietnam, Hong Kong and the flourishing economic zones surrounding it. Its most important responsibility however is the South China Sea and the disputed islands. The PLAAF's Airborne Forces are based in the same area, but is directly assigned to the HQ in Beijing.

Major combat aviation units currently assigned to the Southern Theater Command Air Force (STC AF, 南部战区空军) can be seen on the following pages.

A H-6K under maintenance. This type was delivered to the 24th AR in mid-2012 and there are reports that the regiment moved from Leiyang to Yangtang Li in 2013.
(Top.81 Forum)

The few HU-6 tankers in PLAAF service are all operated by the 23rd AR.
(FYJS Forum)

8th Bomber Division

The 8th Bomber Division was established with Soviet aid in December 1950, at Siping, in Jilin Province, and equipped with Tu-2 bombers. It was based at both Datong and Wenshui prior to 1999, but after 1 October 1999 it was merged with the 48th Bomber Division from the Guangzhou MRAF and has been based at Leiyang since 2003. The status of this important division was long unclear, though it was understood to have been reduced to two regiments, when the original 23rd AR was disbanded in August 1985. It gained again a third regiment in July 1999, when the former 48th Division, 143rd AR was reassigned to the 8th Bomber Division – since 1992 as an independent regiment – and finally renumber to 23rd AR in 2002. Also, several bases have been associated with units of this division. In mid-2012 the 24th AR started receiving the first operational H-6K and in 2015 the 24th AR followed suit replacing the H-6H. Since then this division has operated three regiments, two flying the latest H-6K bombers and the 23rd operating all HU-6 tanker.

Following its sister regiment, the 22nd AR received its H-6K bombers in 2015. This division now only flies this latest variant of the bomber. The unknown store under the wing is likely a KD-63 training round.
(CDF)

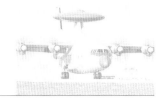

20th Specialised Division

Not much is known about this division, which was originally a bomber division operating Il-2 and later Il-28 and H-5 since 15 September 1951 until 30 September 1992, when it was disbanded. The current 20th Division however can trace it lineage back to the 30th AR, which was first used to establish the 48th Division and later reformed as an independent electronic warfare regiment, before being re-equipped with a mix of specialised Y-8 variants and reassigned to the 10th Bomber Division in 2003. In 2012 the 30th AR moved from Dajiaochang to Guiyang/Leizhuang and became the re-established 20th Division starting from 2013 on as a specialised or special mission division. At around the same time the new 60th AR was formed as a reconnaissance regiment.

A Y-8G (GX-3) flying an ECM mission. This is one of the modified aircraft featuring dark grey painted 'cheeks' and EW antenna fairings.
(Glassy-blue via CDF)

Operated in parallel to the Y-8G within the 58th AR, the Y-8CB (GX-1) is an early ELINT type.
(CJDBY Forum)

Y-8Gs (GX-3) are flown by both the 58th and 59th AR, like this example from the 59th. Altogether eight were built, but the 59th AR lost one (30513) in early January 2018.
(CMA)

Kunming Base

Also in line with the second round of establishing bases and brigades, the Kunming base was formed as the second base within the STC. Today, this base is the successor to the former 44th Fighter Division, which was established in July 1969. Kunming base so far has only three brigades under its control.

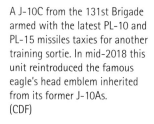

Eagle's head of the 131st Brigade

130th, 131st and 132nd Brigades

These three brigades all stem historically from the 44th Fighter Division, which is one of the least-known PLAAF units and originally equipped with J-6 fighters before converting to J-7s. It was reduced in size to only two regiments somewhere prior to 2012, but regained a third regiment in 2012, when the former Chengdu MR Training Base was merged as the 132nd AR. In that form for many years it was operating three regiments of J-7s, but the 131st AR became the first to receive J-10 fighters, on 13 July 2004. In 2016-17 the 130th AR received J-10A from the 131st AR and the 131st gained new J-10C briefly before both ARs became the 130th and 131st Brigades. In 2014 the 132nd AR relocated from Luliang to Xiangyun and in 2017 it became the 132nd Brigade still acting as a training unit.

A J-10C from the 131st Brigade armed with the latest PL-10 and PL-15 missiles taxies for another training sortie. In mid-2018 this unit reintroduced the famous eagle's head emblem inherited from its former J-10As.
(CDF)

The 132nd Brigade was confirmed as a J-7B/JJ-7A operator only in January 2018 although it was said to have moved from Luliang to Xiangyun as long ago as 2014.
(CDF)

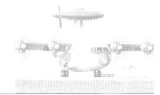

Nanning Base

Under the 2012 introduction of the brigade structure within the PLAAF, the former 42nd Fighter Division and the Guangzhou Military Training Base were unified under the command of Nanning Base, with its HQ at Nanning Wuxu, and associated regiments were expanded to brigades. Again in 2017 the Nanning Base was expanded and reorganised by gaining several more brigades stemming from the former 2nd, 9th and 18th Divisions.

4th, 5th and 6th Brigades

These three brigades all stem historically from the 2nd Fighter Division, which was established with Soviet assistance from the former 11th Regiment of the 4th Combined Brigade on 25 November 1950, initially responsible for air defence of Shanghai, but also took part in the Korean War. In September 1968 the 2nd Fighter Division was relocated from Shanghai to its present location in Guangdong Province, and in the mid-1990s became only the second PLAAF unit to convert to Russian-made Su-27 fighters. Concerning the 4th Brigade, it was originally the 9th Div/27th Reg, which was subordinated to the 2nd Division as the 4th AR. It flew J-8DH and J-8DF decorated with an eagle as tail art and in 2017 the 4th AR was converted to the 4th Brigade. In that turn it replaced its J-8D fighters with the J-11A from the 6th Brigade. The former 5th AR has operated the J-10A/AS since 2006 but replaced this type in November 2015 with J-10Bs. The former 6th AR operated J-11As for many years before the first Su-35s were delivered in December 2016 together with a few Su-30MKKs to serve as trainers for the newly established 6th Brigade.

4th Brigade

The 4th Brigade was long a J-8DH operator and only gained these J-11As in 2017, when the 6th Brigade received its first Su-35. Characteristic for this unit is the huge Chinese flag under the cockpit.
(Tang Jun via chinamil.com)

One of the main missions for the Su-35 is to provide long-range escort for the H-6K cruise-missile-carrying bombers patrolling over the South China Sea.
(Shao Jing via chinamil.com)

The 5th Brigade became the first operator of the improved J-10B in 2015 within the then 5th AR, 2nd Division. However, they are rarely photographed.
(CDF)

26th Brigade

The 26th Brigade is the sole confirmed surviving regiment from the former 9th Fighter Division. Established in December 1950 at Jilin, the original 9th Division was disestablished in September 1955 at the same base and its aircraft transferred to the 5th Division Naval Aviation in December 1955.

A new 9th Fighter Division came into being in March 1956 in Guangzhou, but thereafter very little is known about the activities of this unit. Until 2012 it was a standard division operating three regiments equipped with J-7B, later J-7D and J-7E and J-8D. As noted, the 27th AR was transferred to the 2nd Division as the new 4th AR, the 25th AR disbanded and the remaining 26th AR became the 26th Brigade in 2017. Quite noteworthy the 26th AR was formerly the 35th Division, 103rd AR, which was disbanded in December 2009 before the unit gained J-10A and AS.

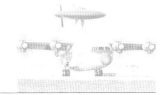

Still unclear – and eventually disbanded in line of the 2017 reorganisation – is the status of the former 25th and the 27th ARs. The 25th was long based at Shaoguan and flew until in 2005 it moved to Shantou and gained J-7E.

The 27th AR was originally the former 30th Division, 90th AR, before it was introduced into the 9th Division as the 27th AR. The unit aircraft initially remained at Pulandian – which is located in the Northern Theater Command – while the 90th AR was converted from the former 12th AR.

A J-10A assigned to the 26th Brigade after a training mission during the Aviadarts exercise, which took place in early August 2018 at central Russia's Ryazan air base.
(Giovanni Colla)

This Su-30MKK assigned to the 54th Brigade shows its famous lightning bolt below an eagle's head, which was introduced in April 2016. The aircraft carries YJ-91ARMs and the new KG600/700 ECM pod.
(CDF)

54th Brigade

Similar to the 26th Brigade, also the 54th is the sole surviving regiment from the former 18th Fighter Division. Established in May 1951 at Guangzhou, the 18th Fighter Division is known to have served at least one tour of duty during the Korean War, and was subsequently involved in operations against the Nationalists while these were still regularly operating over the mainland. In the mid-1960s, it was redeployed to southeast China where it remained until today. It regiments flew J-7s from the 1980s until 2003, when one was disbanded and another converted to Su-30MKKs. In 2012 the division gained again a third regiment from the former Military Region Training Base (MRTB), also equipped with J-7Bs. Quite interesting, the 54th AR received Su-30MKK in 2003 but due to its serial numbers was long thought to be 53rd AR. Finally in March 2017 it traded base and aircraft with the true 53rd AR shortly before being converted to a brigade in June 2017. The two other regiments – the 52nd and 53rd are expected to be disbanded. The Su-30MKK unit is unique in several ways since some aircraft are wearing a green-tan/brown-tan camouflage scheme for DACT and the unit carries a prominent eagle patch under the cockpit.

124th, 125th and 126th Brigades

These three brigades were already established in the first set in 2012, when the former 42nd Fighter Division was disbanded. It was established in July 1969, the division had quite an interesting history gaining its regiments from several other units: the former 125th AR was originally the 18th Division's 53th AR before August 1986, which was established in January 1954. Its units were originally equipped with J-6s, it was re-equipped with J-7s during the late 1980s and consisted of three regiments until the 2012 reorganisations, in which one of its regiments was disbanded. The two remaining regiments flew some of the oldest J-7Hs still active and the division regained its third regiment when the former Guangzhou Military Training Base was merged. Equipment wise two brigades were modernised in recent years, the 124th gained J-10A in late 2013 and in 2016 the 126th replaced its J-7IIs with JH-7A.

The 124th Brigade was established in 2012 as a former J-7H unit assigned to the former 42nd Division. It has been operating J-10A fighters since the end of 2013. (CDF)

Similar to the 124th and 125th Brigades, the 126th Brigade was also a long-time J-7H operator. It replaced these old fighter-bombers with JH-7As in September 2016; this example carries a huge KG800 ECM pod. (Top.81 Forum)

Hong Kong Garrison

Not much is known about the gestation of this new unit. According to reports, the 34th Transport Division's 101st Air Regiment was tasked with the formation of a dedicated helicopter regiment to be based in Hong Kong following the handover of the special administrative region back to the PRC. As such, this Hong Kong Air Regiment was established in March 1993 and probably moved to Sek Kong air base immediately after Hong Kong passed to Chinese sovereignty on 1 July 1997.

Operated by an independent regiment directly assigned to the Hong Kong Garrison, the Z-9ZH (left) is a variant of the Z-9WZ whereas the Z-8KH (right) is comparable to the Z-8KA. The latter features a unique tropical colour scheme. (Left: CMA, right: CDF)

Southern Theater Command HQ Flight and SAR Brigade

Similar to the ETC, also the STC has a dedicated Theater Command transport and search and rescue (SAR) brigade since 2017. This one was the original MRAF HQ flight based at Guangzhou/East. It reports directly to the Southern TCAF Headquarters.

Aviation units assigned to the Southern Theater Command

Code	Unit (Division/Regiment)	Base	Aircraft type	Remarks
	8th Bomber Division			**HQ Leiyang**
18x9x (01-49)	22nd Air Regiment	Shaodong	H-6K	Air base also known as Shaoyang
10x9x (51-99)	23rd Air Regiment	Leiyang	HU-6	
11x9x (00-50)	24th Air Regiment	Leiyang	H-6K	Yantang Li/Xingning is also mentioned as air base
	20th Specialialised Division			**HQ Guiyang-Leizhuangx**
30x1x (01-49)	58th Air Regiment	Guiyang-Leizhuang	Y-8CB (GX1), Y-8G (GX3)	Plus a detachment at Jiaxing
30x1x (51-99)	59th Air Regiment	Luzhou-Lantian	Y-8C, Y-8CB (GX1), Y-8G (GX3)	Based within the WTC
30x1x (51-99)	59th Air Regiment (Det.)	Zunyi-Xinzhou	Y-8CB (GX1), Y-8G (GX3)	
31x1x (01-49)	60th Air Regiment	Zunyi-Xinzhou	JZ-8F, JJ-7A	Status unconfirmed
31x1x (01-49)	60th Air Regiment – Psychological Warfare Squadron (Det.)	Guiyang-Leizhuang	Y-8C, Y-8XZ (GX7)	
	Kunming Base			**HQ Kunming**
74x1x	130th Air Brigade	Mengzi	J-10A/AS	Former 44th AD/130th AR, 50x5x (01-49); received in 2017 J-10A from 131st AR
74x2x	131st Air Brigade	Luliang	J-10B/C, J-10AS	Former 44th AD/131st AR, 50x5x (51-99); wore an eagle patch on the tail and under the cockpit
74x3x	132nd Air Brigade	Xiangyun	J-7H, JJ-7A	Former 44th AD/131st AR, 51x5x (01-49); Gongzhuling is also mentioned as air base
	Nanning Base			**HQ Nanning Wuxu**
61x5x	4th Air Brigade	Foshan	J-11A, Su-27SK/UBK	Former 2nd AD/4th AR, 10x5x (01-49); gained in 2018 J-11A/Su-27UBK from 6th Brigade; wore long time an eagle badge on the tail; most likely relocated in 2017
61x6x	5th Air Brigade	Guilin	J-10B, J-10AS	Former 2nd AD/5th AR, 10x6x (51-99); air base also known as Lijiacun
61x7x	6th Air Brigade	Suixi	Su-30MKK, Su-35	Former 2nd AD/6th AR, 11x7x (01-49)
20x0x (51-99)	26th Air Brigade	Huizhou-Huiyang	J-10A/AS	Former 9th AD/26th AR, 11x7x (01-49);
66x5x	54th Air Brigade	Changsha/City	Su-30MKK	Former 18th AD/53rd AR, 21x9x (01-49); air base also known as Changsha/Datuopu; traded base and aircraft with 53rd AR in March 2017; wearing an eagle patch under the cockpit, some wearing a green-tan/brown-tan camouflage scheme for DACT
73x5x	124th Brigade	Bose/Tianyang	J-10A/AS	
73x6x	125th Brigade	Nanning-Wuxu	J-7H, JJ-7A, Z-8KA?	Said to convert to J-8H from 4th Brigade; Z-8KA most likely transferred to SAR brigade
73x7x	126th Brigade	Liuzhou/Bailian	JH-7A	

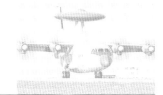

	Hong Kong Garrison			Hong Kong/Shek Kong
6x0x	Independent Helicopter Regiment	Hong Kong/ Shek Kong	Z-8KH, Z-9ZH/WH	Subordinated directly under the CMC

	STC HQ Flight			Hong Kong/Shek Kong
52x1x	Southern Theater Command Transport & SAR Brigade	Guangzhou/East	Mi-17, Mi-171V-5, Y-7G, Z-9	Formerly 6x5x

No forward operational bases are known.

Key

●   Theatre Command HQ

●   PLAAF Command Post

○   Former Command Post

●   PLAAF Air Base

Combat radii

✈ JH-7A   900km (586nm)

✈ J-10A/AS   1,300km (702nm) with tanks

✈ J-11A/B, J-16, 1,340km (724nm) Su-30MKK

✈ H-6K   2,500km (1,350nm)

PLAAF Bases
I   Nanning
II   Kunming
1   Foshan
2   Guilin Li Chan Tsun
3   Suixi
4   Shaodong
5   Leiyang
6   Huizhou–Huiyang
7   Shantou/Waisha
8   Yangtan Li
9   Wuhan/Shanpo (HQ)
10   Changsha City
11   Bose/Tianyang
12   Nanning/Wuxu
13   Liuzhou/Bailian
14   Hong Kong/Shek Kong
15   Guangzhou-East
16   Luzhou-Lantian
17   Mengzi (HQ)
18   Luliang
19   Guiyang/Leizhuang
20   Xiangyun
21   Zunyi-Xinzhou

Islands Key

●   China

●   Philippines

●   Vietnam

●   Malaysia

●   Taiwan

---   Claimed by China

---   Claimed by Philippines

---   Claimed by Malaysia

---   Claimed by Vietnam

---   Claimed by Indonesia

---   Claimed by Brunei

▶

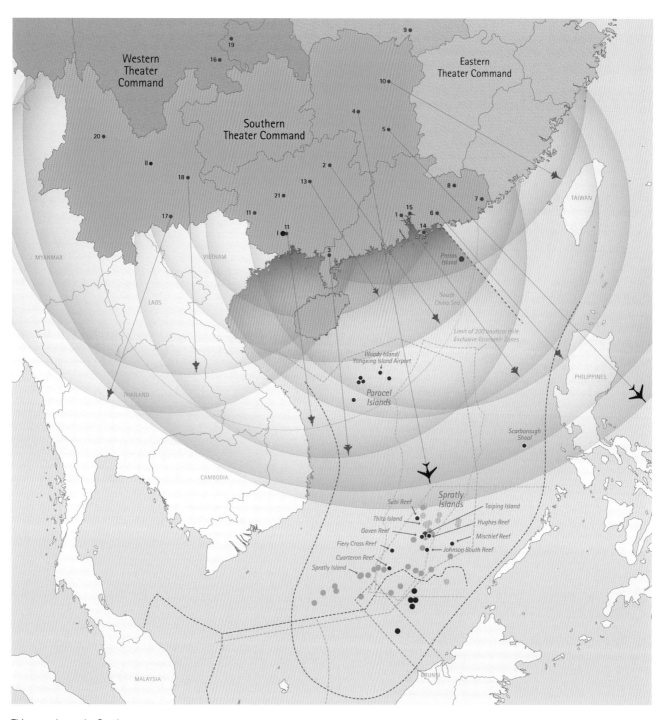

This map shows the Southern
Theater Command and the disputed
islands in the South China Sea. Also
included are the combat radii of the
relevant combat aircraft from their
home bases. The full key can be
found overleaf, on p171. ◀
(Map by James Lawrence)

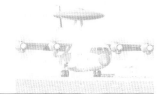

Western Theater Command

The successor to the former Lanzhou Military Region Air Force (Lanzhou jungu kongjun/Lankong) and Chengdu MRAF (Chengdu jungu kongjun/Chengkong), the current Western Theater Command and its WTCAF were founded on 1 February 2016. The Lanzhou MRAF was originally established in May 1952, through the merger of units from the 6th Army and the Xibei Military Region Air Division. Re-designated as the Xibei MRAF (north-western China) from May 1954 until May 1955, this MRAF has been headquartered in Xi'an since late 1952. Its original area of responsibility included the provinces of Sichuan and Tibet, but these were handed over to the Chengdu MRAF in 1965 and 1969, respectively. Currently, the Lanzhou MRAF was responsible for the air defence of the provinces of Gansu, Qinghai and Shaanxi, as well as the autonomous regions of Ningxia Hui, Qinghai, Xinjiang Uyghur as well as the Ngari Prefecture (usually assigned to the Tibet Autonomous Region [AR] and as such the Chengdu MR). The MRAF exercised control over subordinated units via CPs in Hetian, Wulumuqi and Xi'an. The military importance of this MRAF might appear limited at first sight, since it covers the most sparsely populated parts of China. However, it is precisely here that some of China's most secretive military facilities can be found, foremost the Lop Nor nuclear research site ('Base 21') and various missile and EW test facilities. As a result of this fact, combined with the important industrial centre of Xi'an, and proximity to the border with India and the former Soviet Union, the region has been protected since the 1950s.

The history of the Chengdu MRAF can be traced back to January 1950, when the Xinan Military Region Air Division (southwest China) was established in Chongqing. In September of the same year, while planning its advance into Tibet, the PLAAF established the Xinan MRAF, but this was disbanded in June 1955, and its forces transferred to the Lanzhou MRAF. During the 1960s and 1970s the locally deployed PLAAF forces went through a series of reorganisations. The Kunming MRAF Command Post was established in August 1960, and the Chengdu MRAF Command Post in October 1965. These two units were reorganised as the 5th and 8th Air Corps, respectively, several years later, but retained these designations only until November 1978, when they were once again re-designated as the Kunming MRAF CP and the Chengdu MRAF CP.

The Y-9 entered service within the PLAAF in 2012 and is so far operated only by the 10th AR. Production is continuing and the PLA Army Aviation also operates a handful of the type. (FYJS Forum)

Finally, in September 1985, the Kunming MRAC CP and the Chengdu MRAC CP were merged as the Chengdu MRAF. Since all PLAAF forces deployed in the Ngari Prefecture in Eastern Tibet were transferred to the Lanzhou MRAF in 1969, the Chengdu MRAF was responsible for air defence of the provinces of Guizhou, Sichuan and Yunnan, the remaining part of the autonomous regions of Xizang or Tibet and the directly controlled municipality of Chongqing. The Chengdu MRAF exercised control over subordinated units via CPs in Kunming and Lhasa.

In line with the Theater Command reorganisations, the NTC experienced perhaps the most significant changes, since it was established from provinces formerly assigned to three Military Regions: Junan and Shenyang but also large parts of the Beijing MR. Its responsibility encompasses various challenges, such as cross-border terrorism emanating from Central Asia, and the Sino-Indian border. Its jurisdiction includes two reorganised Group Armies (GA): the 76th GA (former 21st GA) and the 77th GA (former 13rd GA). Its joint command headquarters are located in Chengdu. Its WTCAF is therefore now responsible for air defence of the provinces of Gansu, Shaanxi, Sichuan as well as the autonomous regions of Ningxia Hui, Qinghai, Xinjiang Uyghur in addition to Xizang (Tibet) and the directly controlled municipality of Chongqing.

Major units currently assigned to the Western Theater Command Air Force (WTC AF, 西部战区空军) are described on the following pages.

4th Transport Division

This major Transport Division traces its lineage directly back to the first PLAAF unit, the 4th Combined Brigade, which came into being in early 1950, but was reorganised as the 4th Pursuit Brigade in October that year. Initially equipped with MiG-15s and based at Liaoyang in Liaoning Province, it was expanded to become the 4th Fighter Division and served tours of duty during the Korean War. Little is known about its history until the 1980s, when it apparently still flew J-5s. In 2003, while still based at Dalian, the 4th Fighter Division was disbanded and its 10th AR was assigned to the 30th Division as 89th AR. The remaining regiments were then reorganised as a transport division at Qionglai in 2004. During 2009-12 the base at Qionglai was refurbished and expanded. In mid-2012 the 10th AR received the first Y-9 transports and in mid-2016 the 12th AR became the first regiment to introduce the Y-20A.

The Y-20A, which entered service in June 2016, is the pride of the PLAAF's transport units. So far, the only operational unit is the 4th TD, 12th AR. (CDF)

Lanzhou Base

Also in line with the second round of establishing bases and brigades, the Lanzhou base was formed as the second base within the WTC. Today, this base is the successor to the former 33rd Fighter Division, which was established in July 1969. Lanzhou base so far has only four brigades under its control.

16th Brigade

The 16th Brigade is the second surviving regiments formerly assigned to the 6th Fighter Division. However after the original 16th AR was disbanded in 2003 and eventually transformed into a Military Region Training Base (MRTB) flying J-7Bs, the former 47th Division, 140th AR took over the regimental number, which was the final unit to operate the original J-8IE. In line of this transfer the unit reequipped with J-11A until 2002 the and around July 2017 the 16th AR became the 16th Brigade. Quite unique, this unit too shows a prominent unit marking in form of a yellow lion's head.

97th, 98th and 99th Brigades

These two – eventually three – brigades are all former regiments assigned to the 33th Fighter Division. Established at Shanpo, in May 1960, this fighter division flew J-6s during the 1960s and 1970s, but very little is known about its history in general. The former 97th AR gained J-7E from the 111th AR in mid-2011 replacing its J-7Bs and in 2017 it became the 97th Brigade. The 98th Regiment is long operating J-11A and both units wore a prominent winged 33- and eagle's head as its unit markings. In July 2017 it turned into the 98th Brigade and in early 2018 the 98th Brigade became the first PLAAF unit to operate J-16 with low visibility markings. Quite unclear is currently the status of the former 99th AR. Officially flying from Liancheng it was disbanded in 1998 before in April 2012 the former 2nd AR of the Beijing MRTB became the new 99th AR assigned to the 33rd Division. The Beijing MRTB itself was established in October 1988 from the disbanded 17th Division. In line with this change it was reported to have moved from Xishan-Beixiang to Zunyi-Xinzhou, although most sources mention Chongqing/Baishiyi as its base. Today the unit is either disbanded or – as some reports suggest – has replaced its J-7B with either J-7E or even J-7G.

The 97th Brigade is a lesser known unit, confirmed only by a handful of images. It employs unit markings on the J-7E fighters it has operated since 2011.
(www.81.cn)

Carrying the mountain lion as its unit emblem, this J-11A from the 16th Brigade took part in the Sino-Russian anti-terror exercise Peace Mission 2018 in September 2018.
(Anton Harisov)

A similar second-grade unit is the former 33rd Division, 99th AR, which was only confirmed as a brigade in July 2018. Interestingly, its J-7BHs retain the winged eagle's head emblem of the former 33rd Division. (Top.81 Forum)

Lhasa Base

Usually the Theater Commands command two bases each, but as usual in PLA tradition, there's no rule without an exception. Consequently the WTC is so far the sole TC responsible for three bases. Lhasa Base was founded in 2017 but it has not yet any assigned units. Some assume this will only happen in case of forward operational deployments off the bases like Hotan (Hetian Jichang), Kashgar (Kashi or Kaxgar, since mid-2014 the most westerly PLAAF air base), Lhasa/Gonggar, Shigatse (Xigaze East), and Zunyi-Xinzhou. Both Kashgar and Shigatse regularly host UAV deployments have been noted, whereas Lhasa/Gonggar host deployments of J-10, J-11 and KJ-500 since a few years.

In contrast to other bases, Lhasa/Gonggar has no regular units assigned and houses them only during forward operational deployments. In April 2018 it hosted several J-11As (left) from the 16th Brigade and a KJ-500 (right) from the 26th Division. (both Top.81Forum)

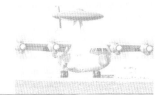

The 98th Brigade was a long-term J-11A operator, before these were replaced in early 2018 with the latest J-16s. With the arrival of these new fighters, low-visibility markings were introduced for the first time and the former prominent yellow unit markings deleted. (SDF)

Ürümqi Base

In 2012 the former 37th Fighter Division was reorganised as the Ürümqi Base and its regiments expanded into brigades. These were expanded in 2017 and reorganised by gaining several more brigades stemming from the former 6th Divisions as well a newly formed UAV unit.

18th Brigade

The 18th Brigade is actually one of two surviving regiments formerly assigned to the 6th Fighter Division. Established in November 1950 at Anshan, in Liaoning Province, this unit completed several tours of duty during the Korean War. Very little is known about its subsequent history, except that it flew J-6s for most of the 1960s and 1970s, receiving F-7s in the 1980s. The unit merged with the deactivated 47th Attack Division in 2003/04 and regained a third regiment by incorporating the former Lanzhou MRTB. In 2017 the 18th AR was transformed into the 18th Brigade and quite unique, this unit shows a prominent red and yellow hawk's head as its unit marking.

109th, 110th and 111th Brigades

These three brigades are all former regiments assigned to the 37th Fighter Division. It was established in August 1966, at Dandong Langtou and was equipped with J-6 and J-7 interceptors for most of its history. However prior to the 2012 modernisations, it emerged as one of the premier PLAAF units. In April 2012 the 37th Division's three regiments were converted into brigades and subordinated to the newly established Ürümqi Base. The latter facility also took command of the former Lanzhou Military Training Base, at Ürümqi South (Wulumuqi). Most likely this was the last PLAAF regiment equipped with the J-6 – eventually the 16th AR – which was disbanded in June 2010. Since mid-2012 the unit began to operate J-7Bs and JJ-7As, as well as a detachment of Y-8 transports at Jiuquan as the 110th Brigade. The 109th has gained J-8H in 2002 replacing J-6 and a few J-8F followed in 2006. The 110th received J-7G in 2004

The 18th Brigade flying J-7Bs is another brigade wearing characteristic markings – in the form of a hawk's head. A J-7B is seen strafing ground targets with unguided rockets during an exercise in early 2018.
(CDF)

One of the major J-8 operators is the 109th Brigade, which introduced J-8Hs in 2002 and additional J-8Fs in 2006.
(CDF)

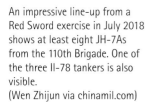

An impressive line-up from a Red Sword exercise in July 2018 shows at least eight JH-7As from the 110th Brigade. One of the three Il-78 tankers is also visible.
(Wen Zhijun via chinamil.com)

replacing J-7B, which were again replaced by JH-7A in 2011. And finally the 111th received J-11B and J-11BS in early 2011 replacing J-7E. Just in April 1989 it has moved from Malan to Korla (Bayingol). Quite surprisingly, in 2016 this brigade also revealed a few examples of the J-11A, which might have replaced the J-11B.

112th Brigade

The 112th Brigade was also formed in 2012 but then still unconfirmed. Concerning its number, the 112th was a regiment formerly assigned to the disbanded 38th until 1988 the regiment operated assigned to the 16th Division based at Ertaizi, which was later transformed into the Shenyang MR Training Base operating JJ-7A since July 2011. In April 2012 the regiment transformed into a brigade operating from Malan.

A rare photo showing a J-11B from the 111th Brigade, taken during an exercise in November 2017.
(CDF)

Xi'an Flight College

Historically, the PLAAF had more than 15 flight schools established, that were later reduced in the 1990s to seven so called flight colleges. In August 2011 the PLAAF implemented a second round of reforms by merging six flight colleges into three. From what is known, the Xi'an Flight College is therefore the successor to the former 2nd Flying School at Huxian and the 5th Flying School at Wuwei. Even if the current status is not entirely clear – especially in mind of where each brigade is located – the Xi'an Flight College seems to be responsible for a later part of the training syllabus since it operates JL-8 basic trainers and JJ-7A advances trainers in at least four training brigades and detachments. Additionally for training helicopter, bomber and transport crews two brigades are equipped with helicopters and Y-7 transports and HYJ-7 bomber trainers.

Together with the J-7B, several JJ-7 and JJ-7A trainers are in use. This is an example from the 1st Training Brigade. The 2nd Training Brigade received its first JL-9 to begin replacing the JJ-7 in mid-2018.
(FYJS Forum)

Besides the JL-8 units, the Xi'an FC also operates two J-7B/JJ-7A units including the 1st Training Brigade – in fact the former Lanzhou MR Training Base – at Jinquan Qingshui, seen here.
(Top.81 Forum)

As one of the three flight colleges, the Xi'an FC operates a huge number of JL-8 trainers in two training brigades including the 3rd Training Brigade at Zhangye/South-East, shown here.
(CDF)

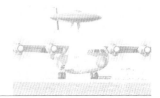

Western Theater Command HQ Flight and SAR Brigade

For the WTC, the dedicated theater command transport and search and rescue (SAR) unit was established in 2017 at Lanzhou. It was originally the former Lanzhou MR Liaison Unit established as an independent transport regiment on 12 April 1967.

However, in line with the 2012 restructuring, it was assigned as the Lanzhou MRAF HQ flight and transformed into a brigade in 2017. It reports directly to the Western TCAF Headquarters but currently it is not entirely clear if it is still based at Lanzhou.

Based on the lastest reports it is officially still a regiment, but its aircraft have already been renumbered and it will soon become a brigade.

Officially still a dedicated transport and SAR regiment, this unit flying Y-7-100 (left) and Y-5C (right) aircraft has already received new serial numbers for the WTC Transport and SAR Brigade.
(**Left:** chinamil.com, **right:** KJ.81.cn)

Together with Lhasa/Gonggar, Shigatse (Xigaze East) has evolved as one of the most important forward operational bases within the WTC. It regularly hosts combat assets from various units – not only from the WTC. Seen here are Su-30MKKs (left) during November 2015 and three Soaring Dragon UAVs (right) during the Sino-Indian Doklam standoff in August and September 2017.
(**Left:** Top.81 Forum, **right:** via East Pendulum)

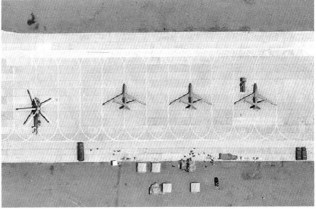

Aviation units assigned to the Western Theater Command

Code	Unit (Division/Regiment)	Base	Aircraft type	Remarks
	4th Transport Division			HQ Qionglai
10x5x (01-49)	10th Air Regiment	Chengdu-Qionglai	Y-8C, Y-9	
10x5x (50-99)	11th Air Regiment	Chengdu-Qionglai	Y-7H	
11x5x (01-49)	12th Air Regiment	Chengdu-Qionglai	Y-20A	
	Lanzhou Base			
62x7x	16th Air Brigade	Yinchuan/West	J-11A, Su-27SK/UBK	Former 6th AD/16th AR, 10x7x (01-49); wearing a lion patch on the tail
70x8x	97th Air Brigade	Dazu	J-7E, JJ-7A	Former 33rd AD/97th AR, 40x4x (01-49); wearing a stylised 97 patch on the fuselage
70x9x	98th Air Brigade	Chongqing-Shashiyi	J-16	Former 33rd AD/98th AR, 10x4x (01-49); air base also known as Baishiyi; wore an eagle patch on the tail
71x0x	99th Air Brigade	Chongqing-Shashiyi	J-7BH, JJ-7A	Former 33rd AD/99th AR, 41x4x (01-49); air base also known as Baishiyi; wears an eagle patch on the tail
	Lhasa Base			
				Founded only in 2017 but not yet with assigned units; most likely used for FODs only
	Ürümqi Base			HQ Ürümqi; also known as Wulumuqi
62x9x	18th Air Brigade	Lintao	J-7H, JJ-7A	Former 6th AD/18th AR, 11x7x (01-49); wearing a hawk's head
72x0x	109th Air Brigade	Changji	J-8F/H, JJ-7A	
72x1x	110th Air Brigade	Ürümqi South	JH-7A	Air base also known as Wulumuqi
72x2x	111th Air Brigade	Korla-Xinjiang	J-11A/B/BS, Su-27UBK	Air base also known as Bayingol; gained again in 2016 some J-11A
72x3x	112th Air Brigade	Malan/Uxxaktal	J-7B, JJ-7A	
	UAV Division			
?	? UAV Battalion	Zhangye/South-East	BZK-005, BZK-006?	
	Xi'an Flight College (former Flying Academy, FA)			HQ Huxian
3x1x	1st Training Brigade	Jinquan Qingshui	J-7B, JJ-7A	Former 2nd TR/2nd FA, 70x2x and Navigator Training School, 72x9x; gained J-7B and JJ-7A from 17th AR
3x1x	1st Training Brigade (Det.)	Huxian	Z-9, EC120?	Status unconfirmed
3x2x	2nd Training Brigade	Hami	J-7B, JJ-7, JL-9	Former 2nd TR/5th FA, 7xx5x
3x3x	3rd Training Brigade	Zhangye/South-East	JL-8	Former 3rd TR/5th FA, 7xx5x
3x4x	4th Training Brigade	Wuwei	JL-8	Former 4th TR/5th FA, 7xx5x

3x5x	5th Training Brigade	Nanchang Gaoping	An-26, Y-7, HYJ-7, Y-8C	Former 3rd TR/2nd FA, 7xx2x

WTC HQ Flight				
53x1x	WTC Transport & SAR Brigade	Lanzhou/ Xiaguanying	Mi-171V-5, Y-5C, Y-7, Z-8K, Z-9B	Former MR HQ base
53x2x	WTC Transport & SAR Brigade (Det.)			

Forward operational bases at: Hotan (Hetian Jichang), Kashgar (Kashi or Kaxgar), Lhasa/Gonggar, Shigatze (Xigaze East) and Zunyi-Xinzhou.

▶

Key		PLAAF Bases				Combat radii		
●	Theatre Command HQ	I	Hetian	8	Changji		JH-7A	900km (586nm)
●	PLAAF Command Post	II	Lanzhou	9	Malan/Uxxaktal			
		III	Lhasa	10	Korla-Xinjiang			
○	Former Command Post	IV	Urumqi (Wulumuqi)	11	Ürümqi South (Wulumuqi)		J-8F/FH, JZ-8F	1,000km (540nm)
		V	Xi'an	12	Kashgar/Kashi			
●	PLAAF Air Base	1	Chengdu-Qionglai	13	Hami			
		2	Dazu	14	Wuwei		J-11A/B, J-16,	1,340km (724nm)
—	Disputed Borders	3	Chongqing-Shashiyi/Baishiyi	15	Lanzhou/Xiaguangying		Su-30MKK	
		4	Lhasa/Gonggar	16	Jinquan Quingshui			
▨	Disputed Territories	5	Yinchuan/West	17	Huxian			
		6	Zhangye/South-East	18	Nanchang Gaoping			
		7	Lintao	19	Shigatse/Xigaze			

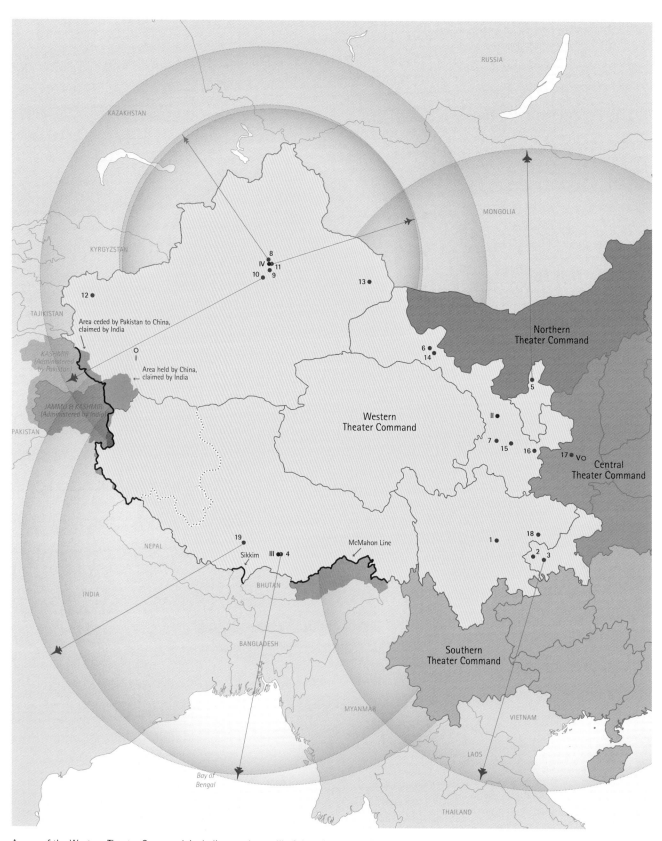

A map of the Western Theater Command, including combat radii of the relevant combat aircraft from their home bases.
The full key can be found overlead, on p183. ◀
(Map by James Lawrence)

Northern Theater Command

The successor to the former Shenyang Military Region Air Force (Shenyang jungu kongjun/Shenkong) and Jinan MRAF (Jinan jungu kongjun/Jikong), the current Northern Theater Command and its NTCAF were founded on 1 February 2016 also inherits major portions of the Beijing Military Region. The Shenyang MRAF was officially established in April 1955. Originally it consisted of the units previously subordinated to the Dongbei Military Region Air Division, established in July 1950, that was renamed the Dongbei MRAF in May 1954. The modern-day composition of the Shenyang MRAF came into being in May 1957, when the Dongbei MRAF was expanded through the inclusion of the Shenyang Military Region Air Defence Force. The Shenyang MRAF was responsible for the air defence of the provinces of Heilongjiang, Jilin and Liaoning and exercises control over subordinated units via the CPs in Changchun and Dalian. This was one of the most important MRAFs within the PLAAF, primarily due to its proximity to Beijing, but also because it is responsible for the protection of such major industrial centres as Harbin and Shenyang, crucial military installations including the Combined Arms Technical Training Base in Changchun, as well as protecting the borders with the Russian Federation and North Korea.

The Jinan MRAF was established in September 1967 from the units of the 6th Air Corps and elements of the Beijing MRAF Headquarters. Initially, it was responsible for air defence of Shandon Province only, but following the disestablishment of the Wuhan MRAF in September 1985, the Jinan MRAF also became responsible for air defence of the Henan Province. Always the smallest of the seven major PLAAF organisations, the Jinan MRAF was also the least well known. It is not known to have any official CP within its structure and its purpose is somewhat unclear: some sources even indicate that the forces assigned to it are considered a strategic reserve.

In line with Theater Command reorganisations, the NTC faced perhaps the most significant changes, since it was established from provinces formerly assigned to three Military Regions: Jinan and Shenyang but also huge parts of the Beijing MR. Its jurisdiction includes the NTC Naval Fleet – the former North Sea Fleet – and three reorganised Group Armies (GA): the 78th GA (former 16th GA), the 79th GA (former 39th GA) and the 80th GA (former 26th GA). Joint command headquarters are located in Shenyang. Its main responsibility is focused on the Korean peninsula and Russia. Its air force is therefore now responsible for air defence of the provinces of Heilongjiang, Jilin, Liaoning and Shandong and it gained responsibility for the Mongolia Autonomous Region (formerly Nei Mongol) but lost Henan, which was transferred to the Central Theater Command.

Major units currently assigned to the Northern Theater Command Air Force (NTC AF, 北部战区空军) are described on the following pages.

Besides the 26th Special Mission Division in the ETC, the PLAAF has established more dedicated EW units. The unit in the NTC is the 16th Specialised Division which operates different Y-8 variants including the Y-8G (GX-3) (left) and the older but rarely seen Y-8CB (GX-1) (right).
(both Top.81 Forum)

The other two regiments assigned to the 16th Specialised Division are the 46th AR flying JZ-8F (left) while the 48th AR operates several smaller transports including this Y-5C (right).
(Left: Top.81-Forum
right: CDF)

16th Specialised Division

Originally established in Qingdao, in Shandong Province, in February 1951, the 16th Division has an eventful history. Initially a fighter unit equipped with MiG-9s and MiG-15s, and subdivided into the 46th and 48th Regiments, it was transformed into the Shenyang Military Region Training Base (MRTB) in August 1988. Slightly later on 26 October 1988 the former 17th Division was reorganised also into the Beijing MRTB. In mid-2012 reports indicated that the former 4th Independent Regiment had been re-formed as the 46th AR and assigned to the re-established specialised division. At around the same time the old JZ-6 were retired and that unit gained JZ-8Fs. The second – also known as the 47th AR – gained special missions Y-8s of different variants as well as a transport flight. Very little is known about the history of the former original 4th Independent Regiment, but it can probably be traced back to the former 4th Independent Bomber Regiment established in December 1952. It was later divided into a separate photo-reconnaissance unit, while the EW unit was equipped with a miscellany of Y-7 and Y-8 variants.

Also it seems as if the 1st AR of the Beijing MR Training Base was merged with the Shenyang HQ transport Regiment to form the new 48th AR at Shenyang-Dongta, however this is unconfirmed.

Operating the J-10A and, since 2007, the J-10AS, the former 1st Division, 2nd AR became a brigade in mid-2017. A J-10AS is seen during a Sino-Russian exercise in August 2017.
(Russian Ministry of Defence)

Dalian Base

Under the 2012 introduction of the brigade structure within the PLAAF, the former 30th Fighter Division and the Shenyang Military Training Base were unified under the command of Dalian Base, with its HQ at Dalian, and associated regiments were expanded to brigades. Again in 2017 the Dalian Base was expanded and reorganised by gaining several more brigades stemming from the former 1st, 11th, 15th and 21st Divisions.

1st, 2nd and 3rd Brigades

These three brigades are all former regiments assigned to the 1st Fighter Division, a division which is considered the 'elite' within the PLAAF. It traces its lineage back to the earliest days of the air force. Originally established as the 10th Regiment within the 4th Combined Brigade, in Nanjing on 16 or 19 June 1950, it was reorganised as the 4th Pursuit or Fighter Brigade, and then as the 4th PLAAF Division, on 31 October 1950. In

The 1st Brigade – in fact the former 1st AR – is a speciality within the PLAAF's J-11B units since it operates a mix of AL-31F-powered single-seaters and WS-10A-powered J-11BS aircraft as seen here.
(FYJS Forum)

March 1956 this unit was deployed to Anshan and was reorganised as the 1st Division, consisting of three regiments. This division served four tours of duty during the Korean War. In 1959, the 1st Fighter Division was one of the first to re-equip with the indigenous J-6 fighter, and ever since has always being the first to introduce the latest and most modern fighters to service. This included the introduction of the J-11B – albeit from Block 01 still with AL-31F engines – which replaced Su-27SK in 2008 in the 1st AR. Similar the 2nd AR replaced their J-7E with J-10A in 2007 and already in 2003 the 3rd AR gained J-8Fs to replace its older J-8Bs in 2003. In mid-2017 all three regiments turned into brigades. Quite interesting, only in late 2017 the 3rd Brigade was merged with the former 62nd AR and most likely moved from Anshan to Qiqihar.

As one of the last J-8H and J-8F units, the current 3rd Brigade is a merged unit comprising fighters from the former 3rd and 62nd ARs. In mid-2018 reports emerged that this unit was due to convert to J-16s later the same year.
(Yang Pan via chinamil.com)

The 31st Brigade transitioned to the JH-7A in 2009 replacing Q-5Ds in then then 31st AR, 11th Ground Attack Division. (CJDBY Forum)

31st and 33rd? Brigades

This brigade is probably the only surviving former regiment from the 11th Ground Attack Division. Established from the former 86th Infantry Division PLA in Xuzhou, in February 1951, the 11th Attack Division was initially equipped with Ilyushin Il-10s, and was subdivided into the 31st and 32nd Attack Regiments. Following at least one tour of duty in Korea, this unit saw involvement in battles with Chinese Nationalists along the eastern coast of the People's Republic of China between November 1954 and August 1958. The unit was reassigned to the Shenyang MRAF in September 1985 and ever since served from local airfields. Following a reorganisation in 1992, the 22nd Attack Division was merged with the 11th Attack Division. Its Q-5Ds were replaced by JH-7As in 2009. The only active unit is confirmed the 31st Brigade, whereas operations for the 32nd AR most likely ended in 2017. The former 33rd AR was already transformed in 2012 to become the 90th Brigade, but there are still reports about a new 33rd Brigade operating J-7s since 2018.

44th Brigade

This brigade stems from the former 15th Ground Attack Division, which was divided in 2017 with the 43rd and 45th remaining within the CTC and the 44th being transferred to the NTC. As a former J-7B regiment assigned originally as the 20th AR, 7th Division, it gained in 2010 J-7G from the Ba Yi demonstration team and in 2012 it was merged into the 15th Division as the new 44th Regiment. In 2017 the 44th Regiment became the 44th Brigade.

61st Brigade

The 61st Brigade is probably the best-known J-10B unit and has been flying the type since 2014. It was involved in the Chinese *Sky Hunter* movie and this particular aircraft is loaded with PL-8B and PL-12 training rounds. (FYJS Forum)

The 61st Brigade is again the sole surviving former regiment once assigned to the 21st Fighter Division. Relatively little is known about this division, except that it was established in November 1951, reportedly at Mudanjiang-Hajlang. Its structure has changed several times since the early 1990s, some of its regiments – such as the 117th – since being disbanded. In 1998 the former 39th Division was merged with the 21st Fighter Division but only in 2003 the 117th AR was renamed into the 62nd AR. This regiment was merged with the 3rd Brigade, the status of the 63rd Brigade is unconfirmed but most likely disbanded and only the 61st became the 61st Brigade in 2017. Since 2014 it operates J-10B replacing J-7Es.

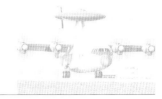

A J-11BS assigned to the 89th Brigade thunders low over a forest during an exercise in July 2018.
(Hao Xiaofeng and Liu Shurui via chinamil.com)

88th, 89th and 90th Brigades

These three brigades were all formed in the first round of reorganisations in 2012 and originate from the former 30th Fighter Division. This unit was established in May 1960 at Donggou and its regiments flew J-6s until re-equipping aircraft taken over from other units, foremost the former 4th Fighter Division, when this was reorganised as a Transport Division. In 2003 the 88th AR started replacing J-8IE with J-7E, and briefly also J-7C/D noted that regiment gained from the former 43rd AR before they were retired in 2015-16. The 89th AR was in fact the former 4th Division's 10th AR, which was subordinated to 30th Division as the 89th AR in 2003. In 2009 the unit moved from Dalian-Zhoushuizi to Pulandian and in 2010 its J-7E were replaced by J-11 with the J-7Es being transferred to the 88th AR. In April 2012 both the 88th and 89th ARs were converted to brigades. The 90th was in fact the former 22nd Division's 66th AR, which was merged in September 1992 with the 11th Division as the new 33rd AR and in 2012 it was reassigned to the Dalian Base as the new 90th Brigade, which reportedly gained new J-7L in 2017-18.

91st Brigade

In line with 2012 modernisation effort, the former base of the 16th and 22nd Divisions was reconstructed as the Shenyang Military Region Training Base (MRTB). One of these training regiments has operated Q-5s and Q-5Js since 2005, while the other has flown JJ-7As since 2011. In 2012 the Regiment was subordinated to the Dalian Base as the newly formed 91st Brigade.

Current status of both the 91st Brigade (left) flying J-7Hs and the 90th Brigade (right) operating Q-5Bs is unconfirmed. Both are said to be either under conversion to more modern types or disbanded.
(Left: Sina,
right: Top.81 Forum)

Jinan Base

Also in line with the second round of establishing bases and brigades, the Jinan base was formed as the second base within the NTC. Today, this base is the successor to the former 12rd Fighter Division but has also one more former 5th Division regiment as a brigade. Jinan Base so far has only three brigades under its control.

15th Brigade

This brigade stems from the former 5th Ground Attack Division, which was established at Kaiyuan, in December 1950. The 5th Attack Division initially consisted of two regiments, both of which completed several tours of duty during the Korean War. Very little is known about its subsequent history except that at some point this division was expanded through the addition of a third regiment, and then equipped with Q-5s. One of these took part in the nuclear test at Lop Nor on 7 January 1972, dropping a live weapon. The 13th and 14th ARs were operating Q-5s until they were withdrawn from use and the regiments disbanded in June 2017, while the 15th AR – which has replaced its Q-5s with JH-7As since 2007 - turned into the 15th Brigade at around the same time.

A JH-7A assigned to the 15th Brigade taxies out for another training mission. This unit gained the type in 2009, replacing the Q-5D. (CJDBY Forum)

34th and 36th Brigades

These brigades were the two remaining former 12th Fighter Division's regiments. This division was established at Xiaoshan in December 1950, and completed at least one tour of duty during the Korean War, possibly also fighting against the Nationalists in the late 1950s. Equipped with J-6s from the mid-1960s, and J-7s from the mid-1980s. Since then its history is somewhat confusing. The 36th AR was based at Jinan air base until 2010, when that air base was abandoned due to urban developments and a new air base at Qihe. In 2012 the 36th gained new J-10As and changed its number plate to the 34th AR. At the same time the old 34th AR became the new 36th and renumbered its J-7Gs accordingly when both units became brigades in 2017. It operates the J-7G since 2012-13 replacing J-7Bs. Quite interesting, the former old 34th AR stems previ-

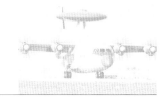

Only two brigades fly the J-7G and both are assigned to the NTC; this one is assigned to the 36th Brigade. The other is the 44th Brigade, which is less well known.
(FYJS Forum)

ously from the Jinan MRTB, which was formed in 1988 from the original 31st Division, 91st AR. Later the Jinan MRTB was merged with the Nanjing MRTB – which in fact was the former 32nd Division, 94th AR before 1988 – as the 34th AR. The status of the 35th AR is unclear, most likely disbanded since it flew the oldest J-8II since 2003.

Unknown/unconfirmed base

As noted in the WTC, Theater Commands usually command only two bases each, but since 2017 the WTC has operated three bases and there are persistent reports of a third base having been established within the NTC somewhere in Inner Mongolia. A possible location is Changchun, as this is a former command post, but this is unconfirmed. If true, this facility is most likely a base for forward operational deployments only – in common with Lhasa Base.

Harbin Flight College

In line with the PLAAF training reform from 2011, the former 1st Flying School (Harbin) and 3rd Flying School (Jinzhou) were reconstituted as the Harbin Basic Flight Training. Again the current status is not entirely clear – especially as regards where each brigade is located – the Harbin Flight College also is responsible for a later part of

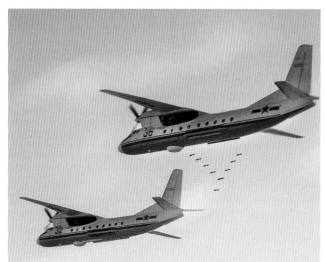

The HYJ-7 bomber trainer (left) features the H-6's bomb sight plus two side bays for several small training bombs. Also part of the Harbin FC are these JL-8 trainers (right) from the 3rd and 4th TBs. The 3rd TB has now converted to the JL-9.
(Left: FYJS Forum,
right: Top.81 Forum)

the training syllabus since it operates even more different types. Currently it flies basic trainers including the CJ-6 and JL-8 in five brigades or detachments, JL-9 advanced trainers in one training brigade and Y-7 transports and HYJ-7 bomber trainers plus a few retired H-6A bombers in two brigades.

In August 2011 the PLAAF announced the formation of two new aerobatic teams. The first of these teams was the Red Falcon Aerial Demonstration Team of the former 3rd Flight Academy, flying JL-8 jet trainers. The second was the Sky Wing Aerial Demonstration Team of the Aviation University, flying CJ-6A trainers.

UAV Brigade?

The oldest and the most modern types in service with the Harbin FC: a handful of H-6As (left) are flown as trainers after being retired as bombers. The latest type is the JL-9 (right), replacing the JL-8 within the 2nd TB.
(Left: KJ.81.cn, right: CMA)

This unit – not yet confirmed as a brigade or division, and perhaps even a battalion – is something of a novelty and its actual status is not fully known. Revealed only in March 2018, it is said to operate a number of dedicated reconnaissance UAVs including the new Soaring Dragon II (EA-03).

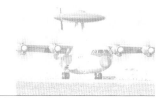

Also assigned to the Harbin FC is the PLAAF's second aerial demonstration formation, the Red Falcon team. It has flown JL-8s since 2011.
(Longshi via CDF)

Northern Theater Command HQ Flight and SAR Brigade

For the NTC, the dedicated theater command transport and search and rescue (SAR) brigade was established in 2017 at Shenyang Dongta.

Its historical lineage is currently unclear, but it might be based on an MR Transport Regiment founded in 2012 and then subordinated to the newly re-established 16th Division as the 48th AR. Another possibility is that it stems from the former 32nd Division's 96th AR. Some sources also mention an additional detachment at Xintai.

Another new unit in PLAAF service is a dedicated reconnaissance brigade or battalion flying Soaring Dragon II UAVs off Shuangliao since March 2018.
(SDF)

Aviation units assigned to the Northern Theater Command

Code	Unit (Division/Regiment)	Base	Aircraft type	Remarks
	16th Specialised Division			HQ Shenyang Dongta
20x7x (01-49)	46th Air Regiment	Shenyang Yu Hung Tun	JZ-8F, JJ-7A	Air base also known as Shenyang-Yuhong
20x7x (51-99)	47th Air Regiment	Shenyang Yu Hung Tun	Y-8C, Y-8CB (GX-1), Y-8G (GX-3)	Might receive KJ-500
21x7x (01-99)	48th Air Regiment	Tongxian	Y-5, Y-7, Z-9WZ	
	Dalian Base			
61x2x	1st Air Brigade	Anshan	J-11B/BS	Former 1st AD/1st AR, 10x2x (01-49); using older AL-31F instead of WS-10A engines
61x3x	2nd Air Brigade	Chifeng	J-10A/AS	Former 1st AD/2nd AR, 10x2x (50-99);
61x4x	3rd Air Brigade	Qiqihar	J-8DH/H/F, JJ-7A, J-16	Former 1st AD/3rd AR, 11x2x (01-49) merged with 21st AD/62nd AR, 30x2x (51-99); moved in 2017 from Anshan to Qiqihar; air base also known as Qiqihar-Sanjiazi
64x2x	31st Air Brigade	Siping	JH-7A	Former 11th AD/31st AR, 20x2x (01-49)
65x5x	44th Air Brigade	Hohhot-Bikeqi	J-7G, JJ-7A	Former 15nd AD/44th AR, 20x6x (51-99)
67x2x	61st Air Brigade	Yanji	J-10B/AS	Former 21st AD/61st AR, 30x2x (01-49); air base also known as Yanji/Chaoyangchuan; most likely relocated in 2018
69x9x	88th Brigade	Dandong/Langtou	J-7E	
70x0x	89th Brigade	Pulandian	J-11B/BS	
70x1x	90th Brigade	Wafangdian	Q-5L/J?, J-7E/L	Reportedly gained J-7E/L in 2017; status unconfirmed
70x2x	91st Brigade	Liuhe	J-7H	Reportedly to gain J-11B/BS
	Jinan Base			
61x6x	15th Air Brigade	Weifang-Weixian	JH-7A	Former 5th AD/15th AR, 11x6x (01-49)
64x5x	34th Air Brigade	Qihe	J-10A/AS	Former 12th AD/34th AR, 20x3x (01-49); some reports mention Gongzhuling Huaide as air base
64x7x	36th Air Brigade	Wendeng	J-7G, JJ-7A	Former 12th AD/36th AR, 20x3x (01-49); some reports mention Gaomi as air base
	Unknown base			In inner Mongolia, possibly Changchun
				Founded only in 2017, but not yet with assigned units; most likely used for FODs only

UAV Brigade?				
?	? UAV Battalion	Shuangliao	BZK-009, Soaring Dragon II	

1xXx	Harbin Flight College (former Harbin Flying Academy, FA)			HQ Harbin
1x1x (?)	1st Training Brigade	Harbin-Shuang Yu Shu	CJ-6, Y-5	Former 1st & 2nd TR/1st FA, 7xx1x; air base also known as Shenyang-Yushu;
1x1x (?)	1st Training Brigade (Det.)	Harbin-Wanggang	CJ-6	Former 1st FA/1st Reg, 7xx1x; merged with 1st FA/2nd Reg; air base also known as Wang Kang;
1x2x	2nd Training Brigade	Harbin-Lalin	HYJ-7, Y-7	Former 3rd TR/1st FA, 7xx1x;
1x2x	2nd Training Brigade (Det.)	Hulun Buir/Hailar	H-6A	Base unconfirmed, but since March 2015 a H-6 unit is based there
1x3x	3rd Training Brigade	Kaiyuan/Tieling	JL-9	Former 2nd TR/3rd FA, 7xx3x; itself former 3rd TR, 7th FA, 7xx7x;
1x4x	4th Training Brigade	Jinzhou-Xiaolingzi	JL-8	Former 3rd TR/3rd FA, 7xx3x;
xx	Red Falcon Aerial Demonstration Team	Jinzhou-Xiaolingzi	JL-8	Established in August 2011
1x5x	5th Training Brigade	Liaoyang	JL-8	Former 4th TR/3rd FA, 7xx3x; Reports also mention Kaiyuan/Tieling as air base

NTC HQ Flight				
54x1x	NTC Transport & SAR Brigade	Shenyang Dongta	Y-5, Y-7, Z-9B, Z-8K	Former 96th Air Regiment, 41x3x (01-49); some sources also mention a SAR brigade or detachment at Xintai

Forward operational base at: Hailar/Southwest

A map of the Northern Theater Command, including the combat radii of the relevant combat aircraft from their home bases. (Map by James Lawrence)

Key

- ⬤ Theatre Command HQ
- ● PLAAF Command Post
- ○ Former Command Post
- ● PLAAF Air Base
- — China ADIZ
- — Japan ADIZ
- — South Korea ADIZ
- — Taiwan ADIZ

PLAAF Air Bases

I	Dalian
II	Jinan
III	Changchun
IV	Shenyang
1	Hohhot-Bikeqi
2	Tongxian
3	Hailar Southwest
4	Anshan
5	Chifeng
6	Kaiyuan/Tieling
7	Wafangdian
8	Shenyang Yu Hung Tun
9	Yanji/Chaoyangchuan
10	Qiqihar
11	Mudanjiang-Hailang
12	Liuhe
13	Pulandian
14	Dandong/Langtou
15	Gongzhuling Huaide
16	Shenyang-Dongta
17	Weifang-Weixian
18	Wendeng
19	Qihe
20	Yanji
21	Shuangliao
22	Jinzhou-Xiaolingzi
23	Siping
24	Liaoyang
25	Harbin/Shuangcheng
26	Harbin/Wanggang
27	Harbin/Shuang Yu Shu
28	Harbin/Lalin
29	Zhuzhen

Combat radii

✈	JH-7A	900km (586nm)
✈	J-8F/FH, JZ-8F	1,000km (540nm)
✈	J-10A/AS	1,300km (702nm) with tanks
✈	J-11A/B, J-16, Su-30MKK	1,340km (724nm)

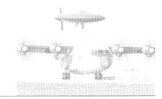

Central Theater Command

Successor to the former Beijing Military Region Air Force (Beijing jungu kongjun/Bei-kong), the current Central Theater Command and its CTCAF were founded on 1 February 2016. The Beijing MRAF was officially established in May 1955, through the reorganisation of the Huabei Military Region Air Division (Northern China), established in 1950 (renamed as the Hubei Air Force Department in May 1954). In common with all other MRAFs of the PLAAF, the Beijing MRAF was expanded through the inclusion of the Beijing Military Region Air Defence Force, in June 1957. The Beijing MRAF was responsible for the air defence of the provinces of Hebei, Neimenggu and Shanxi, the Mongolia Autonomous Region (formerly Nei Mongol), the capital Beijing and the major city of Tianjin, and exercised control over subordinated units via command posts in Datong and Tangshan.

In line with the Theater Command reorganisations, the CTC also faced major changes, since it lost and gained responsibility for several provinces. Its jurisdiction includes three reorganised Group Armies (GA): the 81st GA (former 65th GA), the 82nd GA (former 38th GA) and the 83rd GA (former 54th GA) and its joint command headquarters are located in Beijing. The CTCAF is therefore now responsible for air defence of the provinces of Hebei, Neimenggu and Shanxi, the capital Beijing, and the city of Tianjin and it also acts as a central reserve to provide support to other TCs in case of urgency. Additionally, it gained responsibility for Henan province from the former Jinan MR, the province of Hubei from the former Guangzhou MR but lost responsibility for the Mongolia Autonomous Region (formerly Nei Mongol),which was transferred to the Northern Theater Command. Additional tasks include the training of key personnel for top leadership positions and maintenance of certain crucial air force test facilities.

Major units currently assigned to the Central Theater Command Air Force (CTC AF, (中部战区空军) are described on the following pages.

With the acquisition of additional Il-76MDs and Il-76TDs on the second-hand market, the PLAAF was able to re-equip the 38th AR in 2012-13. The type replaced An-26s and Y-7Hs. This regiment also operates three Il-78 tankers.
(Top.81 Forum)

The Y-8C is the standard workhorse of the PLAAF transport fleet. In particular, the aircraft assigned to the 37th AR at Kaifeng are often used by the Airborne Forces.
(Freeway)

13th Transport Division

The 13th Transport Division traces its heritage back to the PLA's Southwest Military Region Airlift Group, established in April 1950 at Xinjin Xi'an. Originally equipped with Curtiss C-46 and Douglas C-47 aircraft left behind by the Nationalists, in November of the same year the group was expanded through the addition of Soviet-made Ilyushin Il-12 transports and re-designated as the High-Altitude Airlift Regiment. Using personnel transferred from the 2nd Flying School at Changchun, this regiment was finally expanded to become the 13th Transport Division in April 1951. Throughout the 1950s its units were intensively involved in operations against the insurgencies in Tibet and western China, and in the early 1960s in operations against India. In 1962 the 13th Division was assigned to the 15th Airborne Corps. In 1966 the 13th Transport Division was partially re-equipped with Soviet-made An-12 transports, and two years later was reassigned from the 15th Airborne Corps to the Wuhan MRAF, to which it remained subordinated until 1985. Currently, the division operates three regiments; the 37th AR still operates Y-8C transports regularly assigned to the Airborne Troops. The 38th AR flew for long An-26 and Y-7, which were replaced in 2013 by Il-76MD/TDs and three Il-78 tanker. The final regiment, the 39th AR was for long the only one within the PLAAF to operate Il-76MD transports.

For a long time, the 39th AR was the PLAAF's only regiment dedicated to strategic transport duties due to a lack of Il-76MDs in sufficient numbers.
(Top.81 Forum)

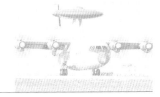

The H-6K bombers assigned to the 108th now appear with correct serial numbers, like this example spotted over the South China Sea in March 2018. In May 2018 bombers from the 108th AR were deployed for the first time to the SCS, arriving at the new base at Yongxing Dao, also known as Woody Island. (Top.81 Forum)

36th Bomber Division

This division was established in March 1965 at Wugong and has operated H-6 bombers ever since. One of its original three regiments – the 106th AR – was disbanded in 2004 most likely to for an independent aerial survey regiment and the other two were still flying H-6. One of them was long said to be equipped with H-6E and H-6Fs and responsible for the nuclear deterrent role, but that was never confirmed. It now flies the H-6H. The final regiment operates a mix of H-6H and updated H-6M – and since January 2017 – also the H-6K. The H-6M are said to go to the re-established 30th AR. Quite interesting and confusing, this division wore off-standard serial numbers for long. Probably since the 106th AR was not a bomber regiment, the 107th AR used numbers usually expected to be used by the 106th and similar the 108th AR used numbers expected for the 107th. This is still valid even after the H-6K were delivered, which now have numbers fitting both the 107th and 108th AR, but are said to be only flown in one regiment. It could however be that both regiments are currently under conversion both in the end applied with correct numbers.

Originally delivered to the 108th AR in January 2017, the first H-6Ks assigned to the 108th had confusing serials that suggested assignment to the 107th AR. Now it seems that both regiments have converted or are under conversion to this latest H-6 variant. This example was seen in March 2018. (FYJS Forum)

Datong Base

When the PLAAF introduced the base/brigade concept in 2012 the former Beijing MR and also the Jinan MR saw no own bases established. This changed now in line of the second round of establishing bases and brigades and the reorganisation to the Theater Commands, so that the Datong Base was formed as the first base within the CTC. Today, this base is the successor to the former 7th and 24th Fighter Divisions but has also one more former 15th Division regiment as a brigade.

19th and 21st Brigades

These two – maybe even a third – brigades are all former 7th Fighter Division's regiments. Established in December 1950 at Dongfeng, the 7th FD originally comprised the 19th and 21st Air Regiments and operated MiG-15s, also during several tours of duty during the Korean War. Little is known about its subsequent history except that the 19th AR flew J-7s before it gained J-11B in 2016. The still unconfirmed 20th AR was in fact a Regiment of the Beijing MRTB before October 1988, when the 17th Division was reorganised. In 2012 the regiment was subordinated to the 7th Division as the 20th AR still operating J-7B but since 2017 its status is unconfirmed and most likely disbanded. The final regiment is the 21st AR, which was transformed into the 21st Brigade in 2017 and at around the same time its old J-7Bs were replaced with J-7Ls reassigned from the 7th and 42nd ARs.

A pair of J-11Bs assigned to the 19th Brigade during the Red Sword 2018 exercise in July 2018. This unit is so far unique since its fighters wear prominent red stripes along the fuselage for formation flying. (Yang Jun via chinamil.com)

Another long-time J-7 user is the 21st Brigade, which gained these J-7Ls from the former 7th AR and 42nd AR in 2017. Note the chaff and flare launchers in the J-7's fin.
(Yang Pan via chinamil.com)

43rd Brigade

The 43rd Brigade is one of two remaining former 15th Ground Attack Division's regiments, after the 15th Division was divided in 2017 with the 44th being transferred to the NTC. Once established in May 1951 at Huiade Xi'an, this unit originally comprised the 43rd and 45th Regiments. Very little is known about its subsequent history, except that it participated in the Korean War, and received Q-5s from the former 50th Attack Division during the 2003/04 reorganisations, when it became a ground attack division. In 2012 it was the only PLAAF division known to operate from bases within another MR. The 43rd AR was one of the last units to operate the J-7C/D until February 2012, when it gained J-10A. In 2017 it was transformed into the 43rd Brigade. The status of the 45th AR is currently unconfirmed and most likely disbanded. Established once in August 1985 as the 50th Division, 149th AR that was subordinated to the 8th Bomber Division in May 1986 still as 149th AR. In April 2005 the 149th AR war resubordinated to the 15th Ground Attack Division as the new 45th AR and by 2017 the final remaining Q-5s were withdrawn from use.

The 43rd Brigade was one of the last units to fly the J-7D, in the then 43rd AR, 15th Ground Attack Division. The lineage of this ground attack unit can be seen here in this bomb-armed J-10A strafing ground targets with its gun.
(Yang Pan via chinamil.com)

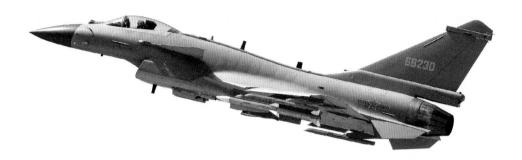

The 72nd Brigade became the second unit to operate the latest J-10C in early 2018, replacing J-10As, which were transferred to the 70th Brigade. (Li Chuan via chinamil.com)

70th, 72nd Brigades and Ba Yi

These two brigades including the Ba Yi – August 1st – PLAAF Aerial Demonstration team are descendants from the former 24th Fighter Division. Established in November 1951 from the cadre of the 70th Dadui at Zunhua, the 24th Fighter Division served at least one tour of duty in Korea, before being reassigned to the Beijing MRAF. During the 1990s all of its regiments were equipped with indigenous J-8 interceptors. The 70th AR replaced these legacy J-8IE with J-7G in 2011 and in 2017 it became the 70th Brigade. In early 2018 they were replaced by J-10A coming from the 72nd Brigade, which converted to J-10C. Concerning the missing 71st AR, there was once a regiment assigned to the Beijing MRTB before it was subordinated to the 33rd Division and renamed into 99th AR. However there are still reports concerning a J-7B Regiment, which is sometimes called the 71st AR or even a brigade, although this is not confirmed. The 72nd, similar to the 70th replaced its J-8IE in 2009 with J-10A, which again were replaced in early 2018 with new J-10C and similar to its sister brigade, the 72nd became a brigade in early 2017.

The Ba Yi or August 1st is the PLAAF's premier aerial demonstration team. In memory of the founding day of the PLAAF on 1 August 1927 it is named 'Ba Yi'. Itself founded

Patch Ba Yi

In 2009 the PLAAF's premier aerial demonstration team '1st August' (Ba Yi) started receiving J-10AY and J-10SY aircraft, as seen here. These are modified variants of the standard J-10A/AS. (Longshi via Top.81 Forum)

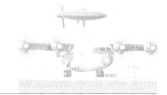

in 1962 and was originally equipped with JJ-5 jet trainers, which were replaced later with J-7EB and in April 2005 with J-7GB. Since 2009 that team operates the dedicated J-10AY and J-10SY variant.

Wuhan Base

Similar to the Datong Base, the Wuhan Base was also established only in 2017 since in 2012 the former Beijing and Jinan MRs had no bases under their command, so that the Datong and Wuhan Bases are now the two bases within the CTC. Today, this base is the successor to the former 19th Fighter Divisions.

55th, 56th and 57th? Brigades

Relatively little is known about the history of the former 19th Fighter Division, except that it was established in November 1951 in Wuhan. For long it had only two regiments but in 2012 one regiment of the former Jinan MRTB – which was formerly from the 31st Division – was added as the new 57th AR. Its current status however is unconfirmed and either disbanded or – as some reports assume – also a brigade now that is said to receive J-11B. The 55th AR gained Su-27SK from the 9th AR at around 2001 and in mid-2017 the 55th AR was transformed into the 55th Brigade. And finally the 56th started somewhere in late 2015 to replace its J-7Bs with J-10Bs before it too was turned into the 56th Brigade in 2017.

This interesting image shows different variants of the Su-27/30 family: on approach in the background an Su-30MKK from the 54th Brigade, to the left a J-11BS, in the middle a J-11B from the 95th Brigade, and finally on the right a rarely seen Su-27SK assigned to the 55th Brigade.
(CJDBY Forum)

Flying the J-10B since 2015, the 56th Brigade was transformed into a brigade in mid-2017. Seen here is a J-10B loaded with PL-12 and the latest PL-10 missiles.
(Top.81 Forum)

Unlike the other two flight colleges, the Shijiazhuang FC only operated JL-8 trainers in four training brigades until 2018. Seen here is an example from the now converted 2nd TB (left), which transitioned to the J-7B and JJ-7A. An interesting shot of the 1st TB's flight line reveals a unit dedicated to training foreign pilots.
(Left: KJ.81.cn,
rigth: FYJS Forum)

Shijiazhuang Flight College

As the two other Flying Academies at Harbin and Xi'an, the Shijiazhuang FC was established in line with the PLAAF training reform from 2011, from the former 4th Flying School (Shijiazhuang) and 6th Flying School (Yongji). As with the other two FAs their current status is not entirely clear – especially in mind of where each brigade is located – the Shijiazhuang Flight College is responsible for an earlier part of the training syllabus since it operates most of all – if not entirely – JL-8 primary trainers. Currently it operates them in four brigades.

A rare image: the PLAAF operates a few An-30s for photo reconnaissance. They are assigned to the re-established Independent Aerial Survey Regiment.
(Sharpgun via FYJS Forum)

Independent Aerial Survey Regiment

The Independent Aerial Survey Regiment is one of the few instances of the PLAAF re-forming a once independent regiment. It was integrated as the 106th AR during the 2012 reform.

Its current assignment is not entirely clear but it is evidently no longer subordinated to the 36th Bomber Division.

Central Theater Command HQ Flight and SAR Brigade

Again akin to the other Theater Commands, the Central TC gained a dedicated theater command transport and search and rescue (SAR) brigade in 2017.

However, very little known about this unit, and its home base is unclear. It reports directly to the Central TCAF Headquarters.

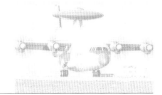

Aviation units assigned to the Central Theater Command

Code	Unit (Division/Regiment)	Base	Aircraft type	Remarks
	13th Transport Division			**HQ Wuhan–Paozhuwan**
20x4x (01-49)	37th Air Regiment	Kaifeng	Y-8C	
20x4x (51-99)	38th Air Regiment	Wuhan-Paozhuwan	Il-76MD/TD, Il-78	
21x4x (01-49)	39th Air Regiment	Dangyang	Il-76MD/TD	Plus possibly a new air base at Mahuiling
	36th Bomber Division			**HQ Lintong**
40x7x (01-49) 40x7x (81-99)	107th Air Regiment	Lintong	H-6H H-6K?	Air base also known as Xi'an/Lintong; H-6H with serial numbers fitting to 106th AR; H-6K have correct serial numbers, but might be renumbered to fit 108th AR
40x7x (51-99) 41x7x (01-29)	108th Air Regiment	Wugong	H-6M H-6K	May move to Neixiang (Ma'ao) where a new air base is under construction; H-6M with serial numbers fitting to 107th AR, but some renumbered correctly; H-6K have correct serial numbers
	Datong Base			
63x0x	19th Air Brigade	Zhangjiakou	J-11B/BS	Former 7th AD/19th AR, 10x8x (01-49); converted to J-11B in 2016
63x1x	21st Air Brigade	Yangging/Yongning	J-7G, JJ-7A	Former 7th AD/21st AR, 11x8x (01-49)
65x4x	43rd Air Brigade	Datong-Huairen	J-10A/AS	Former 15nd AD/43rd AR, 20x6x (01-49)
68x1x	70th Air Brigade	Zunhua	J-10A/AS	Former 24th AD/70th AR, 30x5x (01-49); air base also known as Dongxinzhuang; gained in early 2018 J-10A from 72nd Brigade
68x3x	72nd Air Brigade	Tianjin Yangcun	J-10C/AS	Former 24th AD/72nd AR, 30x5x (51-99); converted in 2018 to J-10C
0x 1x	Ba Yi (1 August) Aerial Demonstration Team	Tianjin Yangcun	J-10AY J-10SY	
	Wuhan Base			
66x6x	55th Air Brigade	Jining	J-11, Su-27SK/UBK	Former 19th AD/55th AR, 30x0x (01-49); based within NTC
66x7x	56th Air Brigade	Zhengzhou	J-10B, J-10AS	Former 19th AD/56th AR, 30x0x (50-99); based within CTC
31x0x (01-49)	57th Air Regiment	Shangqiu	J-7H, JJ-7	Reportedly to gain J-11B/BS; air base also known as Zhuji Guanyintang; based within CTC, status unconfirmed

	Shijiazhuang Flight College			HQ Shijiazhuang Former Shijiazhuang Flying Academy (FA)
2x1x	1st Training Brigade	Shijiazhuang	JL-8	Former 3rd TR/4th FA, coded 7xx4x;
2x2x	2nd Training Brigade	Yongji	J-7B, JJ-7	Former 2nd TR/6th FA, coded 7xx6x;
2x3x	3rd Training Brigade	Xushui-Dingxing	JL-8	Former 3rd TR/6th FA, coded 7xx6x;
2x4x	4th Training Brigade	Tangguantun	JL-8	Former 4th TR/4th FA, coded 7xx4x; air base also known as Jinghai;
	Independent Aerial Survey Regiment			
87x 98x 60xx		Hanzhong- Chenggu	An-30 Y-8H Y-12IV	former 36th BD, 106th Aerial Survey Regiment
	CTC HQ Flight			
55x1x	CTC Transport & SAR Brigade	unkown	Y-7G, Z-8K, Z-9B	

Forward operational bases at: Baoji and Xishanbeixiang (Tong Lin Chuan)

Key

● Theatre Command HQ

● PLAAF Command Post

○ Former Command Post

● PLAAF Air Base

— China ADIZ

— Japan ADIZ

— South Korea ADIZ

— Taiwan ADIZ

PLAAF Air Bases
I Datong
II Wuhan
1 Zhangjiakou
2 Yanging Yongning
3 Datong-Huairen
4 Xushui-Dingxing
5 Zunhua
6 Tianjin Yangcun
7 Tangshan
8 Jining
9 Zhengzhou
10 Hanzhong-Chenggu
11 Lintong
12 Wugong
13 Baoji
14 Yongji
15 Shijiazhuang
16 Xishanbeixiang
17 Tangguantun
18 Xiaogan
19 Wudangshan
20 Dangyang

Combat radii

J-10A/AS   1,300km (702nm)
                  with tanks

J-11A/B, J-16,   1,340km (724nm)
Su-30MKK

H-6K   2,500km (1,350nm)

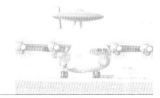

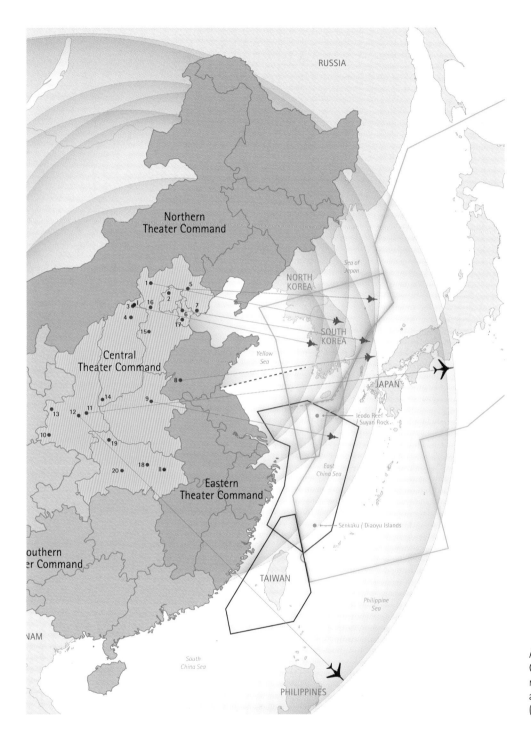

A map of the Central Theater
Command, including combat
radii of the relevant combat
aircraft from their home bases.
(Map by James Lawrence)

PLAAF Headquarters Command/Central Command

Several PLAAF units report directly to the PLA and the PLAAF Headquarters or the Central Command. The best known is the 34th Transport Division specialised in the transportation of senior government and military officials. Because of this, it is often reported as the 'VIP Division'. Others were formerly equipped solely with transport aircraft and some of these are essentially assigned directly to the Army's airborne divisions.

Quite recent additions to the Central Command's assets are the training regiments of both the Aviation University's branches, the Flight Basic Training Base and the Flight Instructor Training Base, which were established from former Flight Academies, and subordinated to the Air Force University at around 2012. Similar, in line with the importance of flight testing and a close cooperation with the mayor Chinese aircraft design and development institutes as well as aircraft manufactures also the Flight Test and Training Base (FTTB) as well as the Chinese Aviation Industry's most important Flight Test and Evaluation facility – the China Flight Test Establishment (CFTE) – are also subordinated under the PLAAF's Headquarters.

And finally, a unit directly subordinated to the PLA General Staff Department is a specialised division operating UAVs, known only as the 'Strategic UAV Scout Force', stationed at a base near Beijing. Its primary equipment consists of BKZ-005 Giant Eagle and Sky Wing UAVs, which may later in the decade be supplemented by the Soaring Dragon, developed since 2000 and used for strategic intelligence gathering operations.

Major units currently directly assigned to the PLAAF HQ are described on the following pages.

34th Transport Division

This unit traces its lineage back to the 34th Independent Transport Regiment, established on 28 August 1963 in Beijing as a unit specialised in the transportation of senior government and military officials. Besides that task, the 34th Division operates also some of the most extensively modified and secretive Tu-154M/Ds and Boeing 737 command posts that remain in PLAAF service. At first, only the 100th and 101st Regiments were established and the 102nd AR founded at Shahe airport followed only in October 1969. In November 1969 it was relocated to Nanyuan following the 13th School had left the air base.

Since then Nanyuan is also homebase for the civil carrier China United Airlines (CUA), which used to be part of the 34th Division. In the meantime more regiments like the 202nd and 203rd were added. Since 2017 it is rumoured however that the 101st, 202nd Regiment and 203rd ARs flying Y-5C, Y-7 and several helicopters were reorganised into a dedicated PLAAF HQ's Transportation and SAR Brigade.

Of the two converted 737-300 airliners acting as command posts, B-4052 was once used as a missile tracking platform for strategic missile tests including the CJ-10 LACM and KD-20 ALCM.
(Zhanghiu via PDF)

Tu-154M B-4014 is one of the few remaining examples serving as airliners. It was photographed at Nanning Wuxu in October 2017.
(Jay Lee)

Complementing the larger A319s and 737s, the PLAAF's 100th AR also uses 10 Bombardier CRJ700. This particular aircraft (c/n 10187) was delivered in May 2005 after being transferred from China United Airlines as C-FCRF.
(Sunshydl)

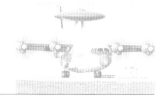

Aviation University | Basic Flight Training Base (BFTB) and Flight Instructor Training Base (FITB)

Beginning in May 2004, the PLAAF founded the Air Force Aviation University in Changchun, by transforming the former 7th Flight College. In August 2011, the PLAAF consolidated six of the previous seven flight colleges and subordinated them to their respective MRAF HQ. These were finally merged into the three current flight colleges. Additionally in April 2012, the former 13th Flight College in Bengbu, was transformed into a Flight Instructor Training Base for flight instructors in the three flight colleges and at operational units, and later a Basic Flight Training Base followed. Both were subordinated in 2017 under the Air Force Aviation University, under direct control of the PLAAF HQ.

In line with this reorganisation, all CJ-6 regiments were concentrated within the then 7th Flying Academy – now forming de facto the core of the BFTB – and all JL-8 regiments were concentrated in the 3rd Flying Academy – by forming the FITB. As such by effectively swapping regiments the training process was streamlined, so that by April 2012 these regiments were subordinated to the Aviation University. In recent years new trainers were introduced: At first the JJ-5 were replaced by JL-8 beginning from 1998 on and since 2014 the JL-9, so that especially the FITB no longer uses JL-8. For basic training so far only the CJ-6 remains in service until a new type is selected and in the future the JL-10 can be expected too.

Flight test and training

Overall the PLAAF has three dedicated flight test and training organisations in service all directly subordinated to the PLAAF Headquarters. These are:

- China Test Flight Establishment – with several Flight Test Regiments (试飞团) at Xi'an-Yanliang
- Subordinate are several Test Flight Flight Groups (试飞大队) at six aircraft factories
- Flight Test and Training Base (飞行试验训练基地) at Dingxin
- Flight Test and Training Base (飞行试验训练基地) at Cangzhou-Cangxian

Whereas the university's Basic Flight Training Base uses almost an entire fleet of CJ-6s, the Flight Instructor Training Base also operates JL-8s – as shown here, from the 1st Training Regiment – and JL-9s. (CDF)

In August 2015 the PLAAF formed its third dedicated aerial demonstration team. Sky Wing is assigned to the PLAAF Aviation University and flies CJ-6A trainers.
(Yang Pan via chinamil.com)

Images of standard trainers assigned to regular training brigades remain rare. This one shows a CJ-6A of the Flight Instructor Training Base in its old green livery with the 7xx8x serial numbers of the former 13th Flight Academy, 3rd Regiment.
(CDF)

Together, both Cangzhou and Dingxin can be seen as two complementary facilities, both based on the difference in size and location. Whereas, Cangzhou focuses on the work to translate new, theoretical guidance on tactics into operational tactics and procedures, Dingxin is located much further away from population centres, thus offering more space for large exercises and giving the PLAAF the opportunity to benefit from live-firing ranges as well as the ability to train under actual electromagnetic jamming conditions.

China Test Flight Establishment (CFTE)

Historically the test centre at Xi'an-Yanliang dates back to the 1950s and is both historically and industrially closely connected to the PRC's aviation industry. The very first

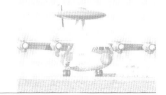

test flight regiment was created in 1959, when the 2nd MID's 4th (Aviation) Bureau formed the China Test Flight Establishment, which is also known as the China Flight Test Research Institute. Sometime after 1984, it was renamed into China Flight Test Research Academy but it always retained the same acronym CFTE. Officially it is known as the China Aviation Industry Test Flight Centre or the AVIC No. 630 Research Institute responsible for all initial testing and the certification of civil and military aircraft, aero engines and airborne equipment.

Since decades the CFTE Test Flight Regiment had at least six test-flight flight groups, including one at each of six major aircraft factories and design institutes. Nearby are the Xi'an Aircraft Industry Group's main production and testing facilities as well as the Xi'an Aircraft Design Institute and the Aerospace and Aeronautical College. Depending the scheduled test, the types of aircraft in use is varying but known are one Il-76LL engine testbed currently testing the WS-20, the KJ-2000 prototype reassigned to WS-18 testing, several H-6 including the latest IFR-equipped variants and examples of each operational combat type plus several others for transport and liaison duties.

As the PLAAF's leading test base, the CFTE runs several important programmes in parallel. Planned to be fitted on the definitive Y-20 is the WS-20 turbofan engine currently under test on an Il-76LL (left). Various updated H-6K prototypes (right) are used for the H-6N project and to test new missiles. (**Left:** Rabbit via FYJS Forum, **right:** CJDBY Forum)

Flight Test and Training Base (FTTB)

Historically the FTTB at Cangzhou – also often known as the Flight Test and Training Centre (FTTC/飞行试验训练中心) – was founded in 1987, when the 11th Aviation School/College at Cangzhou was disbanded. The base itself was constructed in 1953 but due to decades of evolving and expanding it bears no longer any resemblance to its former status. This occurred especially during the late 1980s and early 1990s, when the base solidified its role as the main facility for elite pilot training. Therefore it became home of the first Chinese opposing forces (OPFOR) – also known as Blue Force – unit, equipped with J-10 fighters at a detachment base at Jiugucheng. Reportedly at first dedicated to simulate Soviet air force units first, the OPFOR elements later switched to playing the roles of Taiwan and USAF adversaries. The FTTB maintains a cooperation agreement with Lipetsk air base in Russia, and regularly sends its pilots and controllers there for additional training. The FTTB currently maintains three regiments, also known as Blue Force regiments, which are charged with simulating enemy aircraft in major exercises.

Today the FTTB serves three primary missions: to test any new aircraft under development for operational use under operational conditions, to train the initial cadre of pilots and instructors on that certain type of aircraft before it is deployed to an opera-

Typically, new units are known long before their aircraft are confirmed in images. This case, however, is different: a clear image of a J-7E wearing a number equivalent to the 156th Brigade has been known since 2014, but this unit remains unconfirmed.
(Top.81 Forum)

tional unit and finally to devise new aerial-combat tactics by providing realistic training via the mentioned OPFOR units. Consequently the FTTB operates several dedicated units, namely the 151st, 170th, 171st and 172nd as well as the 177th Brigades:

151st UAV Brigade
This unit is thought to be a relatively new unit rather than an upgraded former FTTB regiment and dedicated for UAV training based at Cangzhou/Cangxian.

156th Air Brigade?
This brigade remains something of a mystery. Traditionally, a unit is first mentioned in a report and its aircraft are only confirmed much later. In this case the situation is different since a very clear image of a J-7E with a serial number indicating assignment to a 156th Brigade is known, but this unit has apparently never been mentioned.

In order to explore the operational use and tactical procedures of UAVs, the PLAAF formed the 151st Brigade as a dedicated UCAV training brigade operating several WD-1Ks, also known as the Wing Loong I.
(Yang Gao)

177th Brigade

170th, 171st, 172nd, 177th and 178th Brigades

These five brigades are all former Cangzhou regimental-level units and in fact respectively they were the former 1st, 2nd, and 3rd FTTB Regiments. The 170th Brigade is thought to be the former 1st or formerly 13th FTTB Regiment based at Jiugucheng originally flying a few J-7E, which were later replaced by J-10A and since 2015-16 by J-10B. The 171st Brigade is thought to be the former Changzhou Dissimilar Air Combat Regiment and may have been the former 2nd FTTB Regiment. The 172nd Air Brigade is thought to be the former 3rd or formerly 14th FTTB Regiment. The final brigade is probably one of the most secretive units: by its serial number it is reportedly not called the 177th but in fact the 66th Brigade. It is a quite new unit formed in 2014 and is the PLAAF's new dedicated OPFOR unit replacing the former small OPFOR unit, which was more the size of a dadui with a true brigade-level unit. Supposedly this unit is to fly two types of aircraft, even if by now only J-11B/BS were noted.

However with much more aircraft operational at Cangzhou than in a standard brigade it is difficult to track the former identity of a certain type. Overall these three brigades are responsible for advanced combat training and the pilots assigned to them are often rated the best within the PLAAF. Quite surprisingly the number 173 and 174 are not allocated.

Historically Dingxin is still regarded as one of the most secretive military facilities in China. Also known as the 14th Air Base or the PLAAF Northwest Tactics Training Base, it is the major testing ground for the development of advanced tactics and techniques required for the effective deployment of air power in combat. The base itself stems from 1958, when the PLAAF built a large base in the Gobi Desert near Dingxin for testing its surface-to-air missiles. During the mid-1990s modernisation, also this base was expanded to act as a flight test and training base. In contrast to the base at Cangzhou however, Dingxin was planned to include a large tactics-training centre where multiple PLAAF units could practice aerial combat developed at operational units and the tactics developed at Cangzhou and various operational units.

Equipment at the facility has varied over time, depending on the requirements of the PLA, ranging from surface-to-surface and surface-to-air missiles, to aircraft equipped with air-to-air missiles. Facilities at Dingxin air base were comprehensively upgraded on three occasions during the 1990s in order to enable the establishment of a Tactical

The 170th Brigade was the first unit to operate the J-10B. It is also the only brigade assigned to the FTTB to operate from Jiugucheng.
(Top.81 Forum)

The 171st Brigade became the first operational JL-9 unit in May 2007 and a batch of five pre-serial JL-9s were tested at the Flight Test and Training Base at Cangzhou for evaluation until 2008. Since then several more have been introduced.
(Top.81 Forum)

As long expected, the JL-10 finally entered service with the FTTB's 172nd Brigade in April 2017. This unit is so far the sole operator.
(B747SPNKG via SDF)

Training Centre, associated with the FTTB. An AEW&C detachment from the CFTE is regularly present at Dingxin to support missile tests.

The nearby Shuangchengzi facility, since renamed as the Jiuguan Space Centre, is one of the PRC's major missile test sites. Also associated with the Dingxin FTTB is Air Force Combined-arms Tactics Training Centre in the Badain Jaran Desert , which was founded in June 1999 and also the PLAAF's 1st, 2nd, and 3rd Test and Training Bases and a Combined-arms Tactics Training Centre, which were approved by the CMC in December 2003. Inaugurated already on 6 January 2004, the mission was identified as merging research, testing, training, and combat under high-tech conditions.

In contrast to Changzhou, the Dingxin Test and Training Base provided the PLAAF with a much larger operating area. The base even includes a mock airfield that appears to be nearly identical to the Taiwan Air Force's Chingchuankang air base in central Taiwan. Consequently it gives the PLAAF unique opportunities to develop proficiency in the strategies and tactics initially developed at Cangzhou.

In addition over the past years, Dingxin became home to 'four key training brands' competitions and exercises including the Golden Helmet competition, the Red Sword exercise, the Golden Dart competition and the Blue Shield exercise (which includes the Golden Shield competition).

In February 2018 the 172nd Brigade at Cangzhou received its first J-20As (left). One of the most secretive units within the FTTB is the 177th Brigade – most likely officially the 66th Brigade. Three of its J-11B/BS together with one naval J-11BSH and the FC-31V1 demonstrator were noted at the SAC factory in Shenyang.
(**Left:** CCTV via CDF,
right: Top.81Forum)

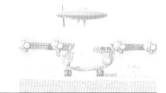

175th and 176th Brigades

These two brigades are permanently based at Dingxin. The 175th Brigade was known until 2012 simply a flight test regiment but before 2005 it was known as the Special Mission Testing Unit (SMTU). In fact it is Dingxin's oldest unit with the mission of testing, trials and exploration. Usually this unit operates a large assortment of aircraft ranging from J-7s, JH-7As to J-11Bs and even Y-8C transports. They are responsible for evaluating weapon systems. This unit is also listed as the 65th Brigade.

The 176th Brigade in contrast is quite a new unit formed in 2014. Its mission is to evaluate new equipment – like the J-10C, J-16 and J-20 – and write the manuals for operational units. It seems as if both brigades are part of the PLAAF's Blue Force.

176th Brigade

The 176th Brigade is an unusual unit since it operates three types: the J-10C, the J-16 shown here and the J-20A. It is the PLAAF's premier operational trials regiment. (haohanfw.com)

The J-20A entered service at the end of 2016 as the PLAAF's first fifth-generation fighter. So far, six are confirmed as operational with the FTTB at Dingxin – part of the 176th Brigade. (Yang Jun via chinamil.com)

Images of the WD-1K UCAVs assigned to the 178th Brigade at Hoxud are rare.
(CDF)

178th Brigade

This brigade is quite a novelty and its actual status is not fully confirmed. By its number it could also be assigned to the flight training brigades similar to the FTTB operated ones, however its base at Malan might be a hint that it is an operational one already. It became the PLAAF's first dedicated UAV brigade operating the WD-1K Wing Loong I UAV/UCAV since 2011.

Strategic UAV Scouting Forces

A further unit directly subordinated to the General Staff Department is a specialised division operating UAVs, known only as the Strategic UAV Scout Force, stationed at a facility close to Shahe air base, near Beijing.

Its primary equipment consists of the BKZ-005 Giant Eagle, which will be complemented by the BZK-009, in development since 2000 and used for strategic intelligence gathering operations. Both UAVs feature a 'stealth'-optimised fuselage design as well as a prominent bulge for a SATCOM antenna above the forward fuselage.

The smaller BZK-005 also has a turret installed underneath the nose, housing various types of cameras.

The BZK-009 is larger and resembles a miniaturised RQ-4 Global Hawk, whereas the even larger Soaring Dragon features a unique 'joined wing' design and is expected to have truly global reach.

The BZK-005 UAV has been in service since the early 2000s, but images of operational aircraft assigned to the Strategic UAV Scouting Force are scarce. The force is not a PLAAF unit, but directly subordinated to the PLA Joint Staff Department. This BZK-005 was on show in a parade in 2015.
(FYJS Forum)

Aviation units directly assigned to the PLAAF Headquarters/Central Command

Code	Unit (Division/Regiment)	Base	Aircraft type	Remarks
	34th Transport Division	HQ Beijing-Nanyuan		Reports directly to Air Force HQ
B-401x			CRJ-200BLR	
B-406x			CRJ-700	
B-409x	100th Air Regiment	Beijing/Xijiao	A319-115 ACJ	
B-400x			737-300	
B-402x			737-700	
B-408x			737-800	
21xx	100th Air Regiment (Det.)	Beijing/Shahezhen	AS332L-1, EC225	
B-407x	101st Air Regiment	Xingtai-Shahe	Y-7	Base unclear, maybe also at Shahe (Hebei)
B-601x			Y-7G	
B-401x			Tu-154M/D	
B-405x	102nd Air Regiment	Beijing-Nanyuan	737-3Q8	
B-408x			LearJet 35A/36A	
6x1x	202nd Air Regiment	Beijing/Shahezhen	Z-8K, Z-9B	Most likely former MR Liaison AR; air base also known as Shahe
6x1x	203rd Air Regiment	Beijing/Shahezhen	Y-5	Most likely former MR Liaison AR;
6x2x			Y-7	air base also known as Shahe
6x3x	??? Air Regiment	Baoji	Y-5	Most likely former MR Liaison AR
4xXx	**Aviation University | Basic Flight Training Base (BFTB)**			**Located in the Northern Theater Command**
4x1x	1st Training Regiment	Changchun-Datun	CJ-6, Y-5	
xx	Sky Wing Aerial Demonstration Team	Changchun-Datun	CJ-6	Reports also mention Jinzhou/Xiaolingzi as air base
4x2x	2nd Training Regiment	Harbin/ Shuangcheng	CJ-6	
4x3x	3rd Training Regiment	Fuxin	CJ-6	
4x4x	4th Training Regiment	Jinzhou/ Liushuibao	CJ-6	
4xXx	**Aviation University | Flight Instructor Training Base (FITB)**			**Located in the Eastern Theater Command**
4x5x	1st Training Regiment	Changchun-Dafangshen	JL-8, JJ-7A	
4x6x	2nd Training Regiment	Zibo	CJ-6	
4x7x	3rd Training Regiment	Yancheng	JL-8	
4x8x	4th Training Regiment	Bengbu	JL-9	Reports also mention Jinzhou/Luishuibao as air base
	China Flight Test Establishment (CFTE)			**HQ Xi'an-Yanliang; also known as Chinese Flight Test Evaluation Centre**
	China Flight Test Establishment (CFTE)	Xi'an-Yanliang	Various types	
	CFTE AEW test facility unit	Dingxin/ 14th Air Base	Various AEW types	
	Flight Test and Training Base (FTTB)			**HQ Cangzhou-Cangxian; also known as 26th Air Base**
76x2x	151st Air Brigade	Cangzhou-Cangxian	WD-1K/GJ-1	The so far first dedicated UCAV unit

76x7x	156st Air Brigade	??	J-7E	Unit unconfirmed, but aircraft identified
78x1x	170th Air Brigade	Jiugucheng	J-10A/AS/B/C, JL-9	
78x2x	171st Air Brigade	Cangzhou-Cangxian	J-7E/JJ-7A, J-8B/D/F, JL-9	Dissimilar Combat OPFOR Regiment.
78x3x	172nd Air Brigade	Cangzhou-Cangxian	Su-30MKK, J-16, JL-10 J-20A	
78x6x	175th Air Brigade	Dingxin/ 14th Air Base	Various: J-7H, JJ-7, J-8F, J-10A/AS/B/C, J-11A/B, J-16, JH-7A, Q-5, Y-5, Y-7, Y-8	Former special mission testing unit, Blue Force unit; also known as Tactical Training Centre (TTC)
78x7x	176th Air Brigade	Dingxin/ 14th Air Base	J-10C, J-16, J-20A	Operational trials regiment
78x8x	177th Air Brigade	Cangzhou-Cangxian	J-11B	Only identified in October 2017; also known as 66th Brigade
78x9x	178th UAV Brigade	Hoxud	WD-1K	Hoxud is a new air base close to Malan
Strategic UAV Scouting Force				
?	Unknown UAV unit	Xingtai-Shahe	CJ-6, BZK-005, BZK-009?	Directly subordinated to the Joint Staff Department, Intelligence Bureau, Information and Communications Bureau

No forward operational bases are known.

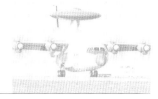

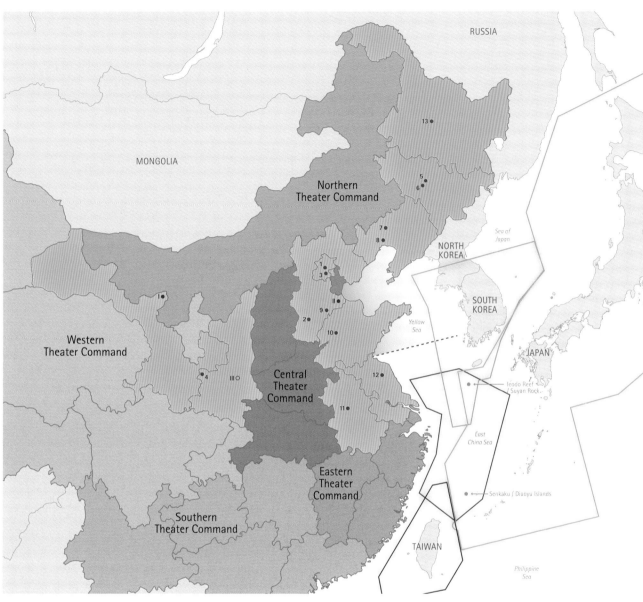

Key

●   Theatre Command HQ

●   PLAAF Command Post

○   Former Command Post

●   PLAAF Air Base

—   China ADIZ

—   Japan ADIZ

—   South Korea ADIZ

—   Taiwan ADIZ

PLAAF Air Bases

I   Dingxin
II   Cangzhou
III   Xi'an
1   Shahezhen (Shahe)
2   Xingtai Shahe
3   Beijing-Nan Yuan
4   Baoji
5   Changchun-Dafangshen
6   Changchun-Datang
7   Fuxin
8   Jinzhou/Liushuibao
9   Jiugucheng
10   Zibo
11   Bengbu
12   Yancheng
13   Harbin/Shuangcheng

A map of the PLAAF Head-quarters/Central Command. (Map by James Lawrence)

PLAAF AIRBORNE CORPS

PEOPLE'S LIBERATION ARMY AIR FORCE AIRBORNE CORPS
(PLAAFAC, 中国人民解放军空降兵军)

As with the PLAN Marine Corps, the reform related to the PLAAF Airborne Forces was published for the first time in the formal announcement by President Xi Jinping in April 2017, when the current overall PLA reforms were instigated. This included not only the reorganisation of the Central Military Commission (CMC), the service headquarters and military regions – to become Theater Commands – but also the re-allocation of personnel and the demobilisation of an unknown number of active duty personnel. These changes are expected to be completed by 2020. Probably the changes will be less dramatic for the Marines and the PLAAF's Airborne Forces although little is known about how far-reaching the modernisation will ultimately be, but, given the Airborne Forces' overall importance within the PLA, their role will not diminish in the near future.

History

For many years, the PLAAF Airborne Corps or Airborne Forces were better known as the 15th Airborne Corps, which has always been the PLA's primary strategic airborne unit. In its current form it is part of the newly-established rapid reaction units (RRU) which exist primarily as airborne and special operations missions. Their role is similar to that of the US Army's 82nd Airborne Division, XVIII Airborne Corps.

Concerning its history, there are still some misconceptions but it can be traced back to 1949, when the Ninth Column was reorganised and pre-designated the 15th Corps. As such, it traces its lineage to an infantry army within the Fourth Field Army – and not, as erroneously reported, from Deng Xiaoping's Second Field Army – to which in fact this corps was later transferred in 1950. As an infantry army it became an airborne troop only on 26 July 1950, when the PLA's 1st Airborne Brigade was implemented for special missions operations. Following the usual Soviet practice, it was at first assigned to the 1st Marine Division, before eventually becoming an airborne division. The brigade's headquarters moved briefly on 1 August to Kaifeng, and on 17 September 1950 the PLA reformed this unit as the PLAAF 1st Airborne Brigade. In the following months it was re-designated several times, becoming also the Air Force Marine 1st

For many years, the Y-5 was the workhorse of the Airborne Forces for liaison, light transport and para jumping. Superseded in most roles by the new Y-12IV, a few dozen Y-5s are still operational. This Y-5B-100 is modified with triple winglets to enhance flight performance. (Chinamil.cn)

Division, the Paratroops Division of the Air Force and finally an Airborne Division. Operationally, it was deployed to Korea in February 1951 and, confusingly, although already subordinated to the Air Force, it fought as part of the 3rd Army Group in April 1951 as an elite unit. In the early 1960s, the 15th Army was reorganised as an airborne army and subordinated to the PLAAF's Headquarters in Beijing. Until the late 1960s the 15th Army's three organic divisions – the 29th, 44th, and 45th – had retained their numerical designations. The 29th Division, however, seems to be an exception: although the reasons why are unclear, it is highly likely that elements of this division merged with the original PLAAF's 1st Airborne Division at Kaifeng to create the 43rd Division. In consequence, in May 1961 the Central Military Commission reassigned the Army's 15th Army into the PLAAF 15th Airborne Army and subordinated the PLAAF's original airborne division to this new Army. Since then all the PLA's paratroop units belong to the PLAAF.

The mid to late 1980s saw several changes: while reduced to three brigades in 1985, in the 1990s the Airborne Corps returned to a divisional structure with an overall increase of 25 per cent in strength. Most notably the percentage of specialised paratroopers rose in order to transform the Airborne Corps into a combined arms force rather than a mobile infantry force. Over the 30 years since their genesis, the Chinese airborne forces developed relatively slowly due to the dogma that the 'army is the most important service'. However, since the late 1990s the emphasis changed dramatically due to two major events. The first was the fast deployment and most successful assault by the US airborne forces in the 1991 Gulf War assisted and/or aided by heavy equipment and secondly, the dramatic earthquake in Sichuan Province in 2008. Since then the PLA has constantly expanded its operational capabilities and combat readiness of its Airborne Forces particularly by intensive, realistic and often international exercises. Prime examples are the Peace Mission joint military exercises carried out since 2005 and the establishment of the new Zhurihe Training Base in northern China's Inner Mongolia Autonomous Region, which was completed in the summer of 2012. That year also saw the initiation of the current reforms to transform the Airborne Forces from a traditional light parachuting 'flying infantry' focused on 'rear combat' to a 'combined air force corps' featuring multiple services, full-time all-domain operation, and the capability of massive airdrops of heavy-duty equipment. But its most fundamental reform occurred in early 2017.

An image from 2013 shows soldiers awaiting to board their Y-7H transports in preparation for a paratrooper exercise. The current status of the Y-7H with the Airborne Forces is unclear. (FYJS Forum)

Current organisation

The Chinese airborne forces are an integral part of the PLAAF and not of the PLA Ground Forces and, although they are nominally part of the PLAAF ORBAT, the 15th Airborne Corps was always under the direct control of the CMC and was not listed within the former Guangzhou MR or current Central Theater Command. The PLA rates its airborne forces not only as an elite unit for special missions operations but also as part of the PLA's strategic reserve. Ever since its transfer to Kaifeng in August 1950 the Chinese Airborne Forces had the Airborne's HQ located at Xiaogan, north of Wuhan in the Henan Province – itself an exception since it was located in the former Jinan MR – with its different divisions, and now brigades, housed at several bases, most of which were within the area of Wuhan in Hubei Province.

In line with the reforms initiated in April 2017, the 15th Airborne Corps became officially the Corps of PLA Airborne Troops or, in short, Airborne Corps. This resulted in the decommissioning of all former divisional headquarters on 19 April and their regiments were re-formed into six airborne brigades (127th, 128th, 130th, 131st, 133rd, and 134th) which were directly assigned under the corps HQ's control. Additionally, a Special Operations, a Strategic Support and an Air Transport Brigade were also formed, so that currently there are a total of nine line brigades. In order to reinforce the concept of being an independent arm of service within the PLAAF the airborne troops deleted the designation of 15th Corps since it carried too strong an imprint of the PLA Army. Besides this structural reorganisation, the commanding systems of the forces have been adjusted accordingly, by transforming the former four-tier commanding system (corps, division, regiment and battalion) to a three-tier system (corps, brigade and battalion) the aim of which is to improve its fast-response capabilities. This 'brigade-centric' organisation mirrors the system that most nations introduced after the end of the Cold War in 1991, in which brigades are now the primary combat units, with their support units permanently attached.

The Y-12IV entered PLAAF service in early 2015 as a successor to the old Y-5 biplane in support of the PLAAF Airborne Forces. They wear a distinctive dark blue/grey camouflage.
(Top.81 Forum)

Mission

The Airborne Corps is used as a principal special operations force employed for independent campaign missions in future wars. This role includes airborne assault landings behind enemy lines in order to paralyse their lines of communication, command and control facilities and to disrupt their movements. It also provides reconnaissance and intelligence-gathering activities, pre-emptive strikes against key military targets including airfields, supply depots and communication centres and also the assassination of key enemy figures. Consequently, the airborne forces would play a major role in any crises with Taiwan or in Tibet and India and could quickly reinforce any forces to restore public order and secure vital military and civil facilities as exemplified during the Cultural Revolution in July 1967 or during the Tiananmen unrests in 1989. In summary, it is a special force in three ways: firstly an offensive one but it is also the ultimate defence force – sometimes the term 'regime preservation force' is used, and finally as a peacetime force in natural disasters and emergencies such as the Sichuan earthquake in 2008. Overall, the corps today is no longer the light paratroop force it once was as it has demonstrated in a number of exercises, also the most often reported analysis that the PLAAF Airborne Force can only deploy one division or equivalent brigades is incorrect. In particular, the increased use of transport planes and helicopters enables it to move a full brigade of paratroops together with their equipment to anywhere in China in under 24 hours. However, given its still limited number of aerial assets, the main mission of these dedicated aviation brigades seems to be training, since they are not big enough to be a major offensive force themselves. They are, however, big enough to constantly train troops in the basics and this is one major difference from the special forces of other nations, which lack individual aerial assets.

The Y-8C is still the largest aircraft directly operated by the Airborne Forces. They were based for many years at Kaifeng together with Y-8Cs assigned to the 13rd TD, 37th AR. However, they likely moved to Yingshan in line with the recent reorganisation.
(Top.81 Forum)

Equipment

Currently, the Chinese airborne force has a strength estimated at 30,000 men (some suggest as many as 35,000), comprising the three former divisions of about 10,000 troops each. It is not only one of the best trained forces within the PLA but is also equipped with the most advanced weapons available to the Chinese military, including modern combat armoured vehicles and infantry gear. This particular has changed in recent times – particularly since 2005 – with the introduction of 'heavy equipment' which includes the addition of portable GPS systems, night-vision goggles, radios and other high-tech equipment as standard fit. Before this, the divisions were equipped to normal infantry division standards apart from the lack of tanks and with only limited artillery, anti-tank, and anti-aircraft weapons. Recently, the inventory has included medium-calibre artillery, towed howitzers and, particularly important, new assault and combat vehicles including the air-dropable infantry fighting vehicle (IFV) ZBD-3 (also known as the ZLC2000), BJ212 jeeps and other 'heavy gear', which is beyond the scope of this chapter. The most important and capable system today is the Z-10K which is used by the force for close air support (CAS).

Of the aerial assets, the most important items of equipment are the transport planes and helicopters. However, bearing in mind that helicopters have been used primarily for transport and liaison duties in the PLAAF and, given that the Army Aviation as a combat force was established only in the late 1980s, it is not surprising that the Airborne Forces were slow to introduce their own aviation units. Consequently, until this branch operated its own aircraft and helicopters, it usually – as with most other nation's airborne troops – requested these systems from other units. For the former 15th Air Corps this was usually the 13th Transport Division with its subordinated regi-

As the PLAAF's primary rapid reaction and special operations force, the Chinese Airborne Forces have demonstrated their ability to move a full brigade of paratroops together with their equipment to anywhere in China in less than 24 hours. (CDF)

ments flying a sizeable number of Y-5s, Y-7s, Il-76MDs and a few helicopters. Since 2009 the Airborne Corps has received more helicopters of its own and the first dedicated Air Corps aviation regiment officially began operations on 3 July 2011 with Z-9s based at Huangpi. In the following years other units received their own types. The latest additions were the Y-12D light STOL transports for basic parachuting and daily parachute training in 2015, the alleged – and still unconfirmed but surprising – phase-out of the Y-7H in 2014 and the equally surprising re-introduction of several Y-5s in 2017. Otherwise, although not directly assigned to the Corps, the PLAAF operates several Il-76MD/MF transports and the most important directly-owned type remains the Y-8C. As for the helicopters, the most important types currently in service are the Z-9WA and improved Z-9WZ, the armed reconnaissance variant Z-9 (about 48), the Z-8KA heavy transport and the specialised SAR helicopters for downed pilots or paratroopers (12) and since late 2015 or early 2016 there have been about 12 Z-10K gunship helicopters.

With regard to the organisational structure, all aerial assets directly operated by the Airborne Corps are assigned to a single air transport brigade – or Transportation Aviation Brigade – which was created in April 2017 and is divided into four separate daduis. In fact, this force has been entirely reorganised similar to a PLA Ground Force Group Army. However, even in July 2018 it the current organisational structure status for the aircraft and helicopters was not clear. The brigade itself is formed of the former 6th Independent Transport Regiment and therefore still carries serial numbers starting with a '6'. Given the relatively small numbers of helicopters and transports, the unit's primary current function seems to be training, whereas, in case of emergency, the airborne troops usually get access to other Army Aviation assets and most importantly the Il-76MF/MD assigned to the 13th Transport Division based at Kaifeng.

The first combat helicopter in Airborne Forces service was the Z-9WA; several improved Z-9WZ were delivered from 2007 onwards. This type is used for battlefield surveillance, armed reconnaissance and –until the Z-10K arrived – for attack missions.
(Top.81 Forum)

Limitations and the future

As noted, the main weakness of the current Airborne Forces is their lack of sufficient airlift capabilities due to the limited number of strategic long-range transports. Although this will undoubtedly be rectified in the future with the steadily increasing numbers of Y-20As and Y-9s, until now the Airborne Corps has been able to deploy only one of its former three divisions – or just two brigades – to anywhere in China within 48 hours. Given the latest reports, besides the Y-20, additional suitable types are possibly the Y-9 and Y-12F as tactical transports to replace the Y-8, Y-5C and Y-12D, the latest Z-8G as a successor for the Z-8KA and, in the longer term, the Z-20. However, questions still remain. The introduction of modern types – particularly the Z-10K attack helicopters and Y-12D for parachute training – has solved many issues but can be only a stopgap, as the numbers in PLAAF service seem too low and these types still lack a satisfactory range. One would expect a greater transport capability for an aviation brigade. Will the Chinese Airborne Corps in the longer term acquire its own long-range assets in greater numbers in order to operate independently without relying on other branches or will they continue to rely on other assets? Will they gain additional support types for EW and/or reconnaissance which are able to act as forward air controllers (FACs) or to assist special forces with heavy EW gear? Will a gunship version of the Y-9 eventually be developed which is comparable to the AC-130 as some observers expect? Finally, it is unclear how the PLA will respond to the newly created Indian Strike Mountain Corps. This can act as a quick reaction force as well as a counteroffensive force against China along the line of control (LAC), and which is intended to reach its full strength of 80,000 troops in the 2020s.

Superseding the Z-9WZ in the attack role, the Z-10K entered service with the PLAAF in early 2016. In contrast to the Army Aviation Z-10A, this variant features new camouflage and several improvements including an enhanced targeting system and the ability to carry different 19-tube 70mm rocket launchers. (CDF)

Conclusion

Li Fengbiao, Chief of Staff of PLA Air Force Airborne Troops, said in 2009: *'We have improved the combat effectiveness of our troops step-by-step. We can perform all-around and rapid mobile operations under many difficult conditions.'* However, in comparison with other countries' airborne forces, the Chinese airborne troops are still inferior in equipment and experience, although they have made remarkable progress in recent years by following the American and Russian training standards both in terms of operations and equipment. In early 2017 the Chinese military site 'China Military Online' posted: *'After over 60 years of development, the Chinese airborne troops have evolved from 'one parachute and one gun per soldier' when it was first established into an 'combined air force corps' featuring multiple services, full-time all-domain operation, and the capability of massive airdrop of heavy-duty equipment.'* For the future, it will be interesting to see how far the airborne troops continue to evolve and if, in the case of the PLAN establishing its own Marine Corps aviation force, both branches will cooperate, follow joint developments or if both see each other as a concurring unit. This situation could become even more confusing if, as possibly intended and in line with recent reforms, each new army group were to receive an assigned army aviation brigade. These could be composed of transport and attack helicopters, and a special operations forces (SOF) brigade, both of which are among the new type of combat units the Army seeks to develop and expand. Consequently, there would be no doubt that the PLAAF Airborne Corps is the PLA's prime force for airborne tasks within and across the Chinese borders and which also plays an ever-increasing role for out-of-area contingencies (OOAC). Since, as the dominant regional power progressing towards becoming a global power, China possesses, and almost certainly will expand, its power projection capabilities in the future.

Aviation units assigned to the People's Liberation Army Air Force Airborne Corps

	Airborne Corps Aviation Brigade	Base	Aircraft type	HQ Xiaogan
6x1x	1st Transport (Fixed Wing) Dadui	Yingshan/ North Guangshui	Y-5, Y-8	Former 6th Transport Regiment, 6x8x (pre-2012) and 6x5x (pre-2017) serial numbers
6x1x	1st Transport (Fixed Wing) Dadui (Det.)	Kaifeng	Y-8C	After the reorganisation, the corps's own Y-8C seem to be relocated to Yingshan
6x1x	2nd Transport (Fixed Wing) Dadui	Xiaogan	Y-7H?	Unconfirmed
6x1x	3rd Transport (Fixed Wing) Dadui	Xiaogan	Y-5, Y-7, Y-12IV	Formerly 6x8x (pre-2012) and 6x5x (pre-2017) serial numbers
6x2x	4th Helicopter (Rotary Wing) Dadui	Huangpi	Z-8KA, Z-9WZ, Z-10K	Formerly 6x6x (pre-2012) and 6x6x (pre-2017) serial numbers

Concerning the correct designation of this aviation brigade, there are still some uncertainties if it is a 'fixed wing aviation brigade' or simply 'aviation brigade' or even 'aircraft brigade' and even more how the different daduis and detachments – especially rotary wing – are called.

A map of air bases used by the
PLAAF Airborne Troops.
(Map by James Lawrence)

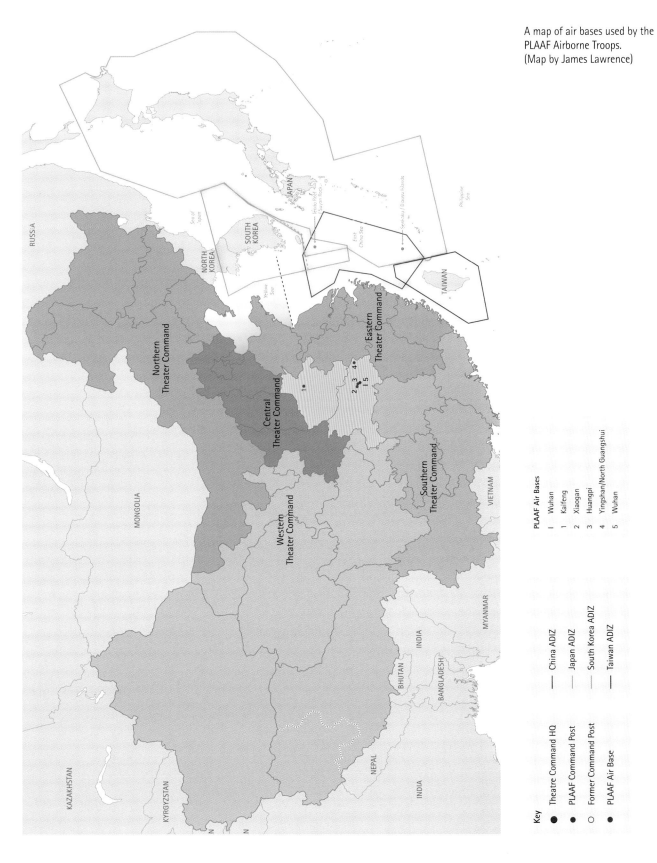

PLAAF Air Bases

I	Wuhan
1	Kaifeng
2	Xiaogan
3	Huangpi
4	Yingshan/North Guangshui
5	Wuhan

Key

●	Theatre Command HQ
●	PLAAF Command Post
○	Former Command Post
●	PLAAF Air Base

——	China ADIZ
——	Japan ADIZ
——	South Korea ADIZ
——	Taiwan ADIZ

中国人民解放军空军

BIBLIOGRAPHY

Allien, J. and Allen, K., 'PLA Air Force's Four Key Training Brands, *CASI*, published 31 May 2018 (http://www.airuniversity.af.mil/CASI/Display/Article/1536348/pla-air-forces-four-key-training-brands/)

Allen, K. W., 'PLA Air Force, Naval Aviation, and Army Aviation Aviator Recruitment, Education, and Training', *Jamestown Foundation*, February 2016 (https://jamestown.org/wp-content/uploads/2016/02/PLA-Aviator-Recruitment-Education-and-Training_Final.pdf)

Blasko, D. J., 'A New PLA Force Structure (Chapter 13), Adjustments in the Command and Control Structure - the Reduction of Military Regions from seven to five', taken from MULVENON, J. C. and YANG, R. H., The People's Liberation Army in the Information Age, (Santa Monica: RAND Corporation, Conference Proceedings, CF-145-CAPP/AF, 1999) ISBN 0-8330-2716-6 (https://www.rand.org/content/dam/rand/pubs/conf_proceedings/CF145/CF145.chap13.pdf)

Blasko, D. J., *The Chinese Army Today: Tradition and Transformation for the 21st Century* (New York: Routledge, 15 February 2012) ISBN 978-04157832-2-4

Blasko, D. J., 'What is Known and Unknown about Changes to the PLA's Ground Combat Units', *China Brief*, Volume 17, Issue 7, 11 May 2017 (https://jamestown.org/program/known-unknown-changes-plas-ground-combatunits/)

Chase, M. S., Allen, K. W., Purser III, B. S., *Overview of People's Liberation Army Air Force 'Elite Pilots'*, (Santa Monica: RAND Corporation; 2016) ISBN: 978-0-8330949-6-4

Clemens, M., 'The Maritime Silk Road and the PLA: Part One', *China Brief*, Volume 15, Issue 6, 19 March 2015 (https://www.cna.org/cna_files/pdf/Maritime-Silk-Road.pdf)

Morris, L. J. and Heginbotham, E., From Theory to Practice: People's Liberation Army Air Force Aviation Training at the Operational Unit, (Santa Monica: RAND Corporation; 2016) ISBN: 978-0-8330-9497-1

Swaine, M. D. and Tellis, A. J., 'Interpreting China's Grand Strategy: Past, Present, and Future', *USNI News*, 26 May 2015 (http://news.usni.org/2015/05/26/document-chinas-military-strategy#NSS).
See also http://www.rand.org/pubs/monograph_reports/MR1121.html

Tao, Zhang (editor), 'China's Military Strategy – The State Council Information Office of the People's Republic of China', *Xinhua*, 26 May 2015 (http://english.chinamil.com.cn/news-channels/2015-05/26/content_6507716.htm)

Trevethan, L., *Brigadization of the PLA Air Force*, (Santa Monica: RAND Corporation/CASI, 1 May 2018) ISBN 978-1718721159

Villasanta, A. D., 'China Strengthening People's Liberation Army Navy Marine Corps at Expense of the Army', *China Topix*, 16 March 2016 (edit 2017) (http://www.chinatopix.com/articles/112482/20170316/china-strengthening-peoples-liberation-army-navy-marine-corps-expense.htm#ixzz4gPhoysL5)

Wuthnow, J. and Saunders, Ph. C., 'Chinese Military Reform in the Age of Xi Jinping: Drivers, Challenges, and Implications', *China Strategic Perspectives*, No. 10, National Defense University Press Washington, D.C.: Center for the Study of Chinese Military Affairs Institute for National Strategic Studies, March 2017 (http://ndupress.ndu.edu/Portals/68/Documents/stratperspective/china/ChinaPerspectives-10.pdf)

Xi Zhigang and Jiang Long, 'In-depth: A close look at Chinese airborne troops', China Military Online, 30 August 2017 (http://eng.chinamil.com.cn/view/2017-08/30/content_7736996.htm)

Further readings

Butowski, P., *Russia's Warplanes, Volume 1, Russian-made Military Aircraft and Helicopters Today* (Houston: Harpia Publishing LLC, 2015) ISBN 978-09854554-5-34

Butowski, P., *Russia's Warplanes, Volume 2, Russian-made Military Aircraft and Helicopters Today* (Houston: Harpia Publishing LLC, 2016) ISBN 978-09973092-0-1

Butowski, P., *Russia's Air-launched Weapons, Russian-made Aircraft Ordnance Today* (Houston: Harpia Publishing LLC, 2017) ISBN 978-09973092-1-8

http://www.chinadaily.com.cn/china/2011-07/13/content_12897211.htm

http://china-pla.blogspot.de/2011/08/evolution-of-plaaf-doctrinetraining.html

https://jamestown.org/wp-content/uploads/2016/02/PLA-Aviator-Recruitment-Education-and-Training_Final.pdf

http://www.rand.org/blog/2016/10/chinas-plaaf-pilot-training-program-undergoes-major.html

http://www.rand.org/pubs/research_reports/RR1415.html

http://www.rand.org/content/dam/rand/pubs/research_reports/RR1400/RR1416/RAND_RR1416.pdf

中国人民解放军空军

INDEX

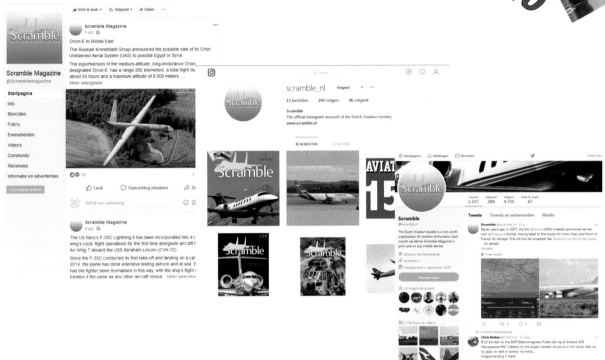